Marvelous Modular Origami

Jasmine Dodecahedron 1 (top) and 3 (bottom). (See pages 50 and 54.)

Marvelous Modular Origami

Meenakshi Mukerji

CRC Press
Taylor & Francis Group
Boca Raton   London   New York

CRC Press is an imprint of the
Taylor & Francis Group, an **informa** business

Editorial, Sales, and Customer Service Office

A K Peters, Ltd.
5 Commonwealth Road, Suite 2C
Natick, MA 01760
www.akpeters.com

Library of Congress Cataloging-in-Publication Data

Mukerji, Meenakshi, 1962–
    Marvelous modular origami / Meenakshi Mukerji.
        p. cm.
    Includes bibliographical references.
    ISBN 978-1-56881-316-5 (alk. paper)
    1. Origami. I. Title.

    TT870.M82 2007
    736′.982--dc22

                                2006052457

ISBN-10 1-56881-316-3

Cover Photographs

Front cover: Poinsettia Floral Ball.

Back cover: Poinsettia Floral Ball (top) and Cosmos Ball Variation (bottom).

Printed in India

14 13 12 11 10   10 9 8 7 6 5 4 3 2

To all who inspired me
and to my parents

Contents

Preface

Never did I imagine that I would end up writing an origami book. Ever since I started exhibiting photos of my origami designs on the Internet, I began to receive innumerable requests from the fans of my website to write a book. What started as a simple desire to share photos of my folding unfolded into the writing of this book. So here I am.

To understand origami, one should start with its definition. As most origami enthusiasts already know, it is based on two Japanese words *oru* (to fold) and *kami* (paper). This ancient art of paper folding started in Japan and China, but origami is now a household word around the world. Everyone has probably folded at least a boat or an airplane in their lifetime. Recently though, origami has come a long way from folding traditional models, modular origami being one of the newest forms of the art.

The origin of modular origami is a little hazy due to the lack of proper documentation. It is generally believed to have begun in the early 1970s with the Sonobe units made by Mitsunobu Sonobe. Six of those units could be assembled into a cube and three of those units could be assembled into a Toshie Takahama Jewel. With one additional crease made to the units, Steve Krimball first formed the 30-unit ball [Alice Gray, "On Modular Origami," *The Origamian* vol. 13, no. 3, June 1976]. This dodecahedral-icosahedral formation, in my opinion, is the most valuable contribution to polyhedral modular origami. Later on Kunihiko Kasahara, Tomoko Fuse, Miyuki Kawamura, Lewis Simon, Bennet Arnstein, Rona Gurkewitz, David Mitchell, and many others made significant contributions to modular origami.

Modular origami, as the name implies, involves assembling several identical modules or units to form a finished model. Modular origami almost always means polyhedral or geometric modular origami although there are a host of other modulars that have nothing to do with polyhedra. Generally speaking, glue is not required, but for some models it is recommended for increased longevity and for some others glue is required to simply hold the units together. The models presented in this book do not require any glue. The symmetry of modular origami models is appealing to almost everyone, especially to those who have a love for polyhedra. As tedious or monotonous as folding the individual units might get, the finished model is always a very satisfying end result—almost like a reward waiting at the end of all the hard work.

Modular origami can fit easily into one's busy schedule. Unlike any other art form, you do not need a long stretch of time at once. Upon mastering a unit (which takes very little time), batches of it can be folded anywhere, anytime, including very short idle-cycles of your life. When the units are all folded, the assembly can also be done slowly over time. This art form can easily trickle into the nooks and crannies of your packed day without jeopardizing anything, and hence it has stuck with me for a long time. Those long waits at the doctor's office or anywhere else and those long rides or flights do not have to be boring anymore. Just carry some paper and diagrams, and you are ready with very little extra baggage.

Cupertino, California
July 2006

Acknowledgments

Photo Credits

So many people have directly or indirectly contributed to the happening of this book that it would be almost next to impossible to thank everybody, but I will try. First of all, I would like to thank my uncle Bireshwar Mukhopadhyay for introducing me to origami as a child and buying me those Robert Harbin books. Thanks to Shobha Prabakar for leading me to the path of rediscovering origami as an adult in its modular form. Thanks to Rosalinda Sanchez for her never-ending inspiration and enthusiasm. Thanks to David Petty for providing constant encouragement and support in so many ways. Thanks to Francis Ow and Rona Gurkewitz for their wonderful correspondence. Thanks to Anne LaVin, Rosana Shapiro, and the Jaiswal family for proofreading and their valuable suggestions. Thanks to the Singhal family for much support. Thanks to Robert Lang for his invaluable guidance in my search for a publisher. Thanks to all who simply said, "go for it", specially the fans of my website. Last but not least, thanks to my family for putting up with all the hours I spent on this book and for so much more. Special thanks to my two sons for naming this book.

Daisy Sonobe Cube (page xvi): photo by Hank Morris

Striped Sonobe Icosahedral Assembly (page xvi): folding and photo by Tripti Singhal

Snow-Capped Sonobe 1 Spiked Pentakis Dodecahedron (page xvi): folding and photo by Rosalinda Sanchez

90-unit dodecahedral assembly of Snow-Capped Sonobe 1 (page 5): folding by Anjali Pemmaraju

Calla Lily Ball (page 16): folding and photo by Halina Rosciszewska-Narloch

Passion Flower Ball (page 28): folding and photo by Rosalinda Sanchez

Petunia Floral Ball (page 28): folding and photo by Carlos Cabrino (Leroy)

All other folding and photos are by the author.

Platonic & Archimedean Solids

Here is a list of polyhedra commonly referenced for origami constructions.

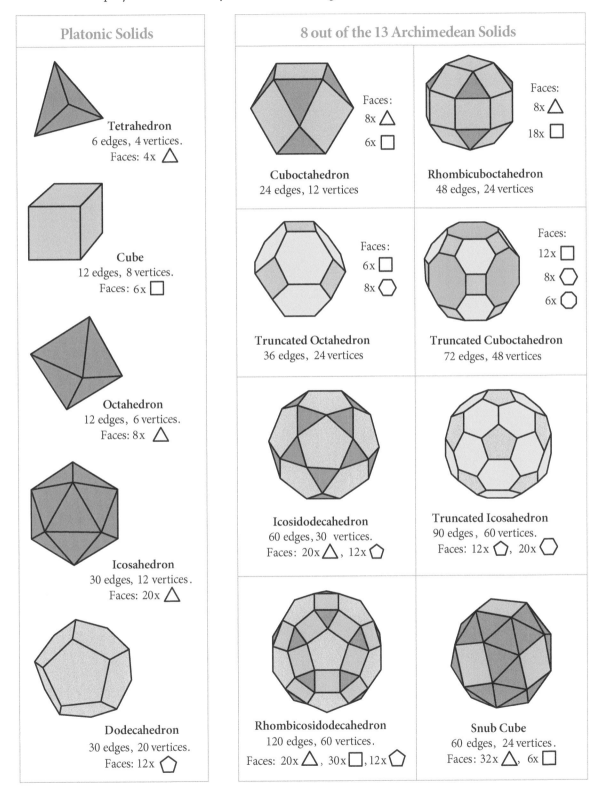

Platonic Solids	8 out of the 13 Archimedean Solids	
Tetrahedron 6 edges, 4 vertices. Faces: 4x △	**Cuboctahedron** 24 edges, 12 vertices Faces: 8x △ 6x □	**Rhombicuboctahedron** 48 edges, 24 vertices Faces: 8x △ 18x □
Cube 12 edges, 8 vertices. Faces: 6x □	**Truncated Octahedron** 36 edges, 24 vertices Faces: 6x □ 8x ⬡	**Truncated Cuboctahedron** 72 edges, 48 vertices Faces: 12x □ 8x ⬡ 6x ⬡
Octahedron 12 edges, 6 vertices. Faces: 8x △	**Icosidodecahedron** 60 edges, 30 vertices. Faces: 20x △, 12x ⬠	**Truncated Icosahedron** 90 edges, 60 vertices. Faces: 12x ⬠, 20x ⬡
Icosahedron 30 edges, 12 vertices. Faces: 20x △		
Dodecahedron 30 edges, 20 vertices. Faces: 12x ⬠	**Rhombicosidodecahedron** 120 edges, 60 vertices. Faces: 20x △, 30x □, 12x ⬠	**Snub Cube** 60 edges, 24 vertices. Faces: 32x △, 6x □

Origami Basics

The following lists only the origami symbols and bases used in this book. It is not by any means a complete list of origami symbols and bases.

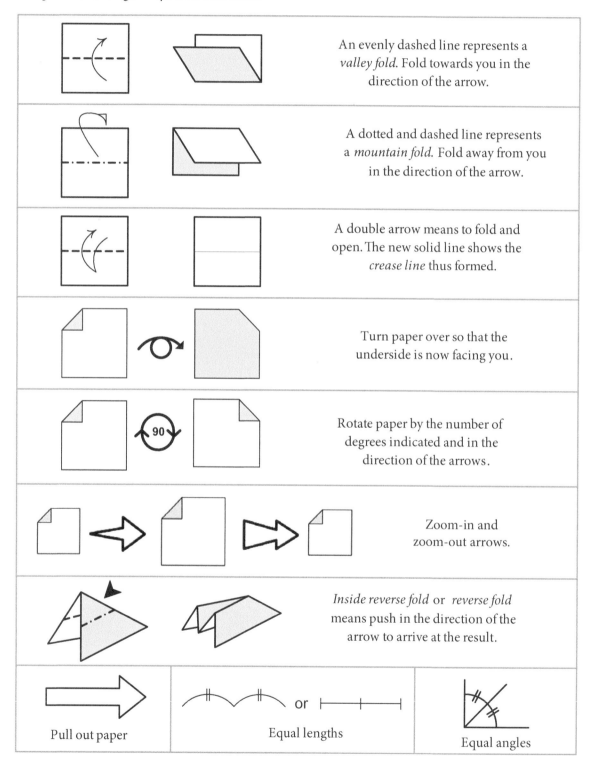

An evenly dashed line represents a *valley fold*. Fold towards you in the direction of the arrow.

A dotted and dashed line represents a *mountain fold*. Fold away from you in the direction of the arrow.

A double arrow means to fold and open. The new solid line shows the *crease line* thus formed.

Turn paper over so that the underside is now facing you.

Rotate paper by the number of degrees indicated and in the direction of the arrows.

Zoom-in and zoom-out arrows.

Inside reverse fold or *reverse fold* means push in the direction of the arrow to arrive at the result.

Pull out paper

Equal lengths

Equal angles

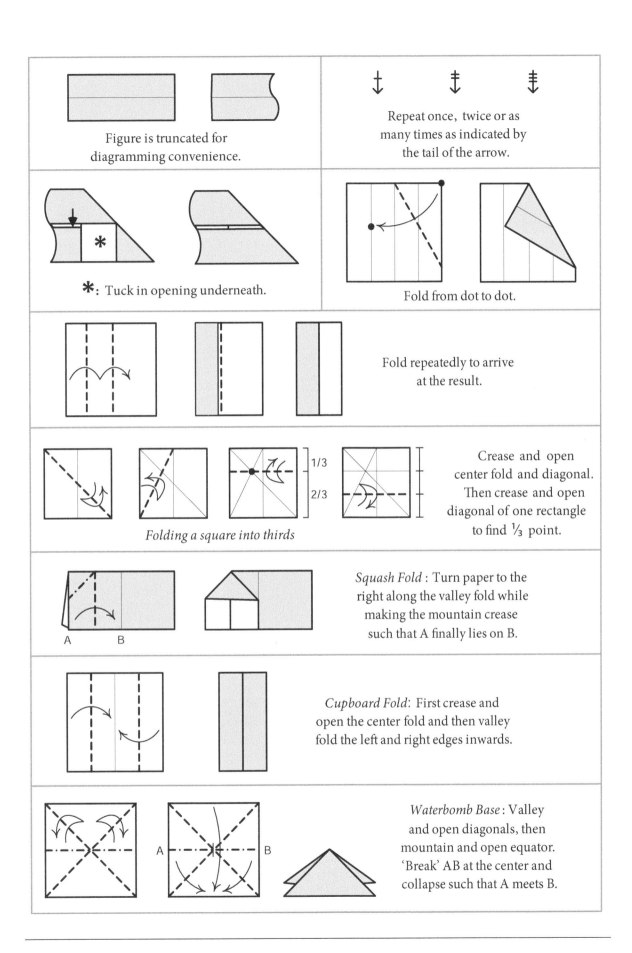

Figure is truncated for diagramming convenience.

Repeat once, twice or as many times as indicated by the tail of the arrow.

*: Tuck in opening underneath.

Fold from dot to dot.

Fold repeatedly to arrive at the result.

Folding a square into thirds

Crease and open center fold and diagonal. Then crease and open diagonal of one rectangle to find ⅓ point.

Squash Fold : Turn paper to the right along the valley fold while making the mountain crease such that A finally lies on B.

Cupboard Fold: First crease and open the center fold and then valley fold the left and right edges inwards.

Waterbomb Base: Valley and open diagonals, then mountain and open equator. 'Break' AB at the center and collapse such that A meets B.

Folding Tips

◈ Use paper of the same thickness and texture for all units. This ensures that the finished model will hold evenly and look symmetrical. One does not necessarily have to use the origami paper available in stores. Virtually any paper from color bond to gift-wrap works.

◈ Pay attention to the grain of the paper. Make sure that, when starting to fold, the grain of the paper is oriented the same way for all units. This is important to ensure uniformity and homogeneity of the model.

◈ It is advisable to fold a trial unit before folding the real units. This gives you an idea of the finished unit size. In some models the finished unit is much smaller than the starting paper size, and in others it is not that much smaller. Making a trial unit will give you an idea of what the size of the finished units and hence a finished model might be, when you start with a certain paper size.

◈ After you have determined your paper size, procure *all* the paper you will need for the model before starting. If you do not have all at the beginning, you may find, as has been my experience, that you are not able to find more paper of the same kind to finish your model.

◈ If a step looks difficult, looking ahead to the next step often helps immensely. This is because the execution of a current step results in what is diagrammed in the next step.

◈ Assembly aids such as miniature clothespins or paper clips are often advisable, especially for beginners. Some assemblies simply need them whether you are a beginner or not. These pins or clips may be removed as the assembly progresses or upon completion of the model.

◈ During assembly, putting together the last few units, especially the very last one can get challenging. During those times remember that it is paper you are working with and not metal! Paper is flexible and can be bent or flexed for ease of assembly.

◈ After completion, hold the model in both hands and compress gently to make sure that all of the tabs are securely and completely inside their corresponding pockets. Finish by working around the ball.

◈ Many units involve folding into thirds. The best way to do this is to make a template using the same size paper as the units. Fold the template into thirds by the method explained in "Origami Basics." Then use the template to crease your units. This saves time and reduces unwanted creases.

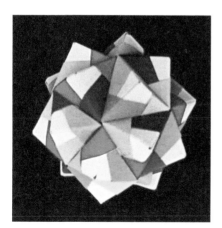

Daisy Sonobe Cube (top left), Striped Sonobe Icosahedral Assembly (top right), Snow-Capped Sonobe 1 Spiked
Pentakis Dodecahedron (middle), and Snow-Capped Sonobe 1 (bottom left) and 2 (bottom right) Icosahedral Assemblies.

1 ◆ Sonobe Variations

As previously discussed in the preface, the Sonobe unit is one of the foundations of modular origami. The variations presented in this chapter may have been independently created by anyone who has played around enough with Sonobe units like I have. Nevertheless, it is worthwhile to present some of my variations in a dedicated chapter.

The Daisy Sonobe is my very first own creation. I borrowed the idea of making variations to simple Sonobe units to achieve dramatic end results from modular origami queen Tomoko Fuse. After mak-ing some of these models, you will be on your way to creating your own variations.

With the addition of extra creases to a finished unit as listed in the table on page 2, Sonobe units can be assembled into a 3-unit Toshie's Jewel, a 6-unit cube, a 12-unit large cube, a 12-unit octahedral assembly, a 30-unit icosahedral assembly, a 90-unit dodeca-hedral assembly, other bigger polyhedral assem-blies, and even other objects such as birds, flowers, and wreaths. You may try making any shape from the table with any Sonobe variation.

Toshie's Jewels made with Sonobe variations (clockwise from top left: Swan Sonobe,
Snow-Capped Sonobe 1, Daisy Sonobe, Snow-Capped Sonobe 2, and Striped Sonobe).

Sonobe Table

Model	Shape	# of Units to Fold	Finished Unit Crease Pattern
Toshie Takahama's Jewel		3	
Cube		6	
Large Cube		12	
Octahedral Assembly		12	
Icosahedral Assembly		30	
Spiked Pentakis Dodecahedral Assembly		60	
Dodecahedral Assembly		90	

Sonobe Assembly Basics

Sonobe assemblies are essentially "pyramidized" polyhedra with each pyramid consisting of three Sonobe units and each unit, in turn, being a part of two adjacent pyramids. The figure below shows a generic Sonobe unit and how to form one pyramid. When constructing a polyhedron, the key thing to remember is that the diagonal *ab* of each Sonobe unit will lie along an edge of the polyhedron.

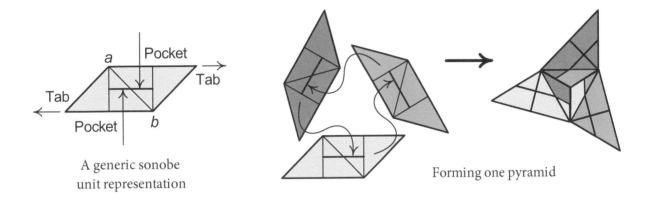

A generic sonobe
unit representation

Forming one pyramid

Sonobe Assembly Guide for a Few Polyhedra

1. Toshie's Jewel: Crease three finished units as explained in the table on page 2. Form a pyramid as above. Then turn the assembly upside down and make another pyramid with the three loose tabs and pockets. This assembly is also sometimes known as a Crane Egg.

2. Cube Assembly: Crease six finished units as explained in the table on page 2.

Each face will be made up of the center square of one unit and the tabs of two other units. Do Steps 1 and 2 to form one face. Do Steps 3 and 4 to form one corner or vertex. Continue interlocking in this manner to arrive at the finished cube.

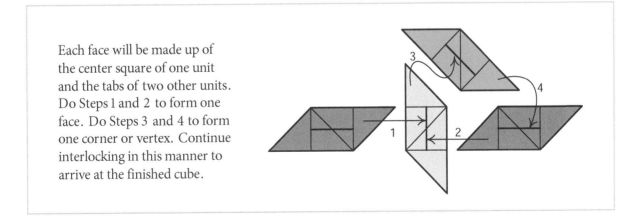

3. Large Cube Assembly: Crease 12 finished units as explained on page 2.

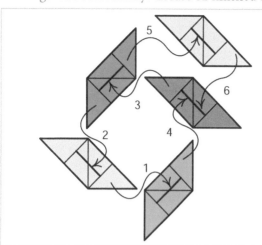

The 12-unit large cube is the only assembly that does not involve pyramidizing. Each face is made up of four units with each unit being a part of two adjacent faces. Do Steps 1–4 to form one face. Do Steps 5 and 6 to form a vertex or corner. Continue forming the faces and vertices similarly to complete the cube.

4. Octahedral Assembly: Crease 12 finished units as explained on page 2.

Assemble four units in a ring as shown following the number sequence. Take a fifth unit and do Steps 5 and 6 to form a pyramid. Continue adding three more units to form a ring of four pyramids. Complete model by forming a total of eight pyramids arranged in an octahedral symmetry.

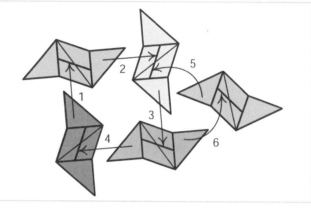

5. Icosahedral Assembly: Crease 30 finished units as explained on page 2.

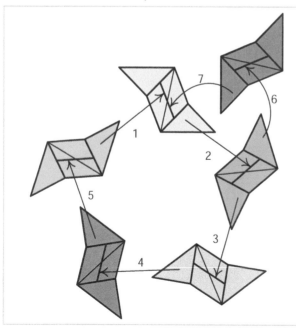

Assemble five units in a ring as shown following sequence numbers. Take a sixth unit and do Steps 6 and 7 to form a pyramid. Continue adding four more units to form a ring of five pyramids. Complete model by forming a total of 20 pyramids arranged in an icosahedral symmetry.

6. **Spiked Pentakis Dodecahedral Assembly:** This model will be discussed at the end of this chapter. Please see page 15.

7. **Dodecahedral Assembly:** This is similar to the icosahedral assembly. Fold 90 units and crease the finished units as explained in the table on page 2. Form a ring of five pyramids. Surround this with five rings of six pyramids such that each of the first five original pyramids is also a part of a ring of sixes. Continue in this manner to complete the ball. You can also think about this assembly as a dodecahedron where the faces are not flat but consist of a ring of five pyramids.

Striped Sonobe Cube, Swan Sonobe Octahedral Assembly, and Daisy Sonobe Large Cube.

90-unit dodecahedral assembly of Snow-Capped Sonobe 1.

Daisy Sonobe

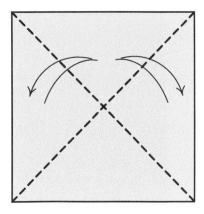

1. Crease and open diagonals.

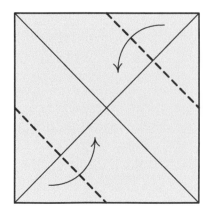

2. Bring two corners to center.

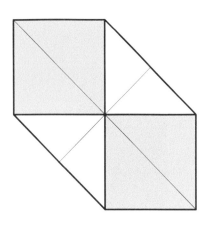

3. Turn over.

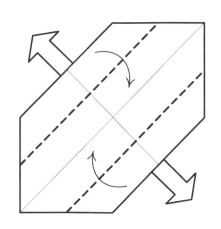

4. Bring edges to center, top layer only.

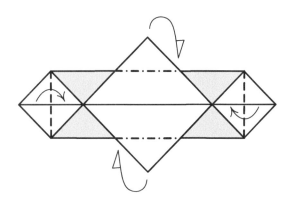

5. Tuck back top and bottom corners under first layer. Fold in left and right corners.

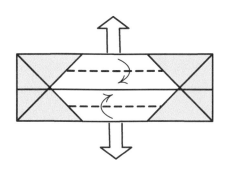

6. Fold as shown, top layer only.

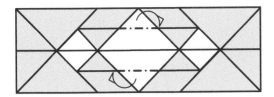

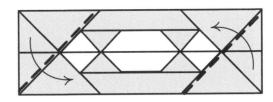

7. Mountain fold corners and tuck back.

8. Valley fold as shown.

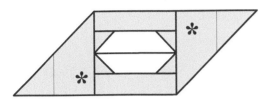

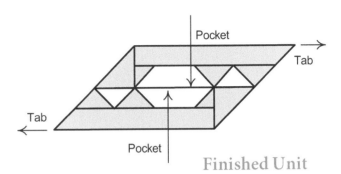

9. Tuck flaps marked ✸ in opening underneath.

Pocket

Tab

Tab

Pocket

Finished Unit

Assembly

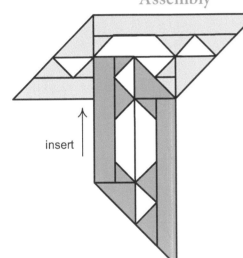

insert

Refer to pages 2–5 to determine how many units to fold, the crease pattern on the finished unit, and how to assemble.

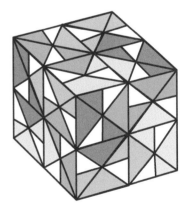

12-unit large cube assembly

6-unit cube assembly

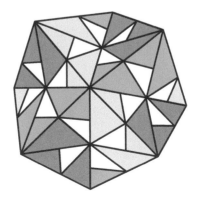

12-unit octahedral assembly

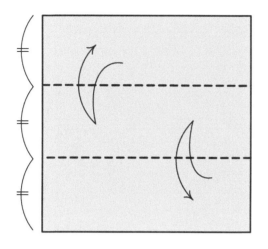

1. Valley fold into thirds and open.

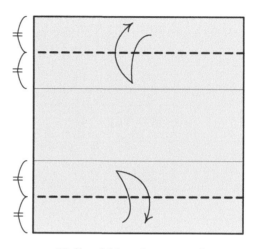

2. Valley fold and open as shown.

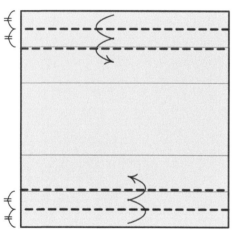

3. Valley fold twice on each edge.

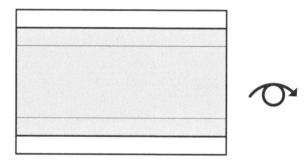

4. Turn over.

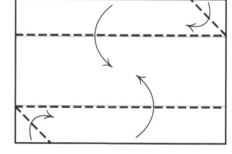

5. Fold corners, then valley fold existing creases.

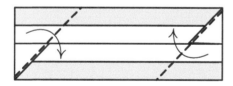

6. Valley fold corners.

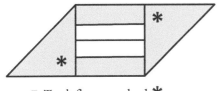

7. Tuck flaps marked ✱
in opening underneath.

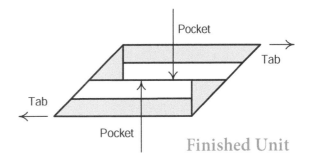

Pocket

Tab

Tab

Pocket

Finished Unit

A Variation:
To get narrower stripes of
white, fold into thirds instead
of halves in Step 3.

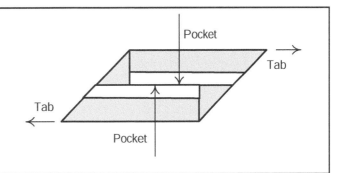

Pocket

Tab

Tab

Pocket

Assembly

Refer to pages 2–5 to determine how
how many units to fold, the crease
pattern on the finished unit, and how
to assemble.

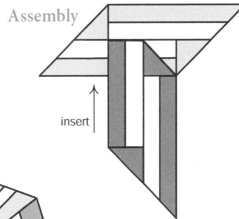

insert

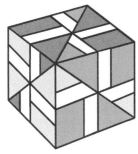

6-unit cube
assembly

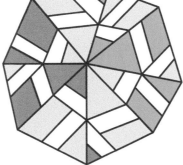

12-unit octahedral
assembly

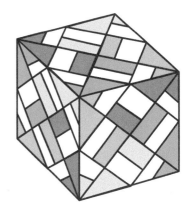

12-unit large
cube assembly

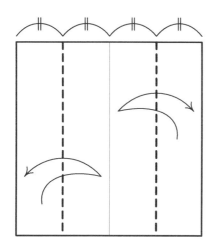

1. Cupboard fold and open.

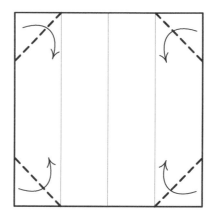

2. Fold corners, then rotate 90°.

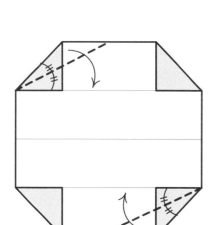

3. Fold corners to bisect
marked angles.

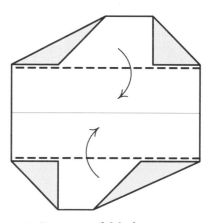

4. Re-crease folds from
Step 1.

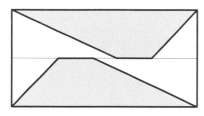

5. Turn over.

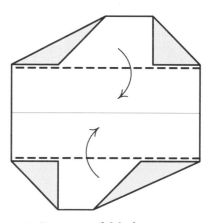

6. Fold corners.

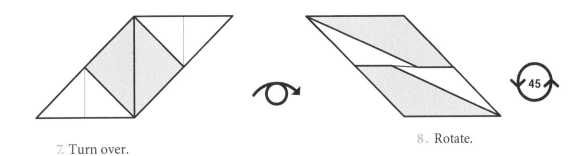

7. Turn over.

8. Rotate.

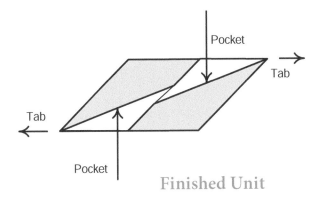

Finished Unit

Refer to pages 2–5 to determine how
many units to fold, the crease pattern
on the finished unit, and how to assemble.

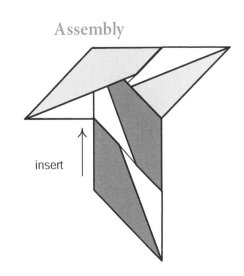

Assembly

insert

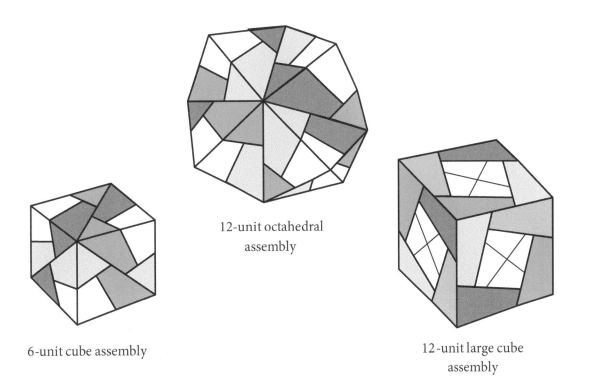

6-unit cube assembly

12-unit octahedral
assembly

12-unit large cube
assembly

Snow-Capped Sonobe 2

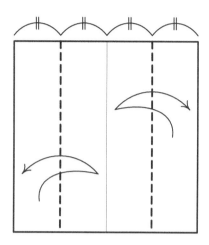

1. Cupboard fold and open.

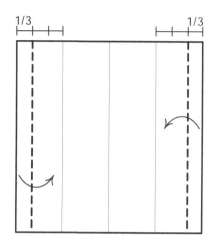

2. Fold thirds as shown.

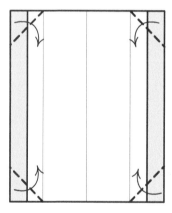

3. Fold corners.

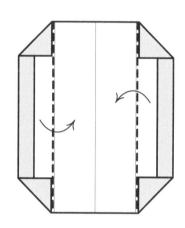

4. Re-crease cupboard folds.

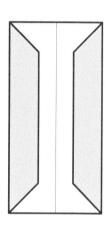

5. Turn over.

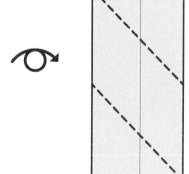

6. Fold corners.

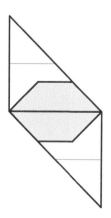

7. Turn over.

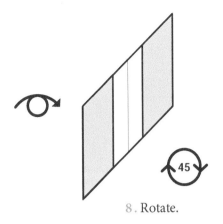

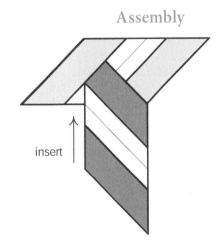

Pocket

Tab

Tab

Pocket

Finished Unit

8. Rotate.

Assembly

Refer to pages 2–5 to determine
how many units to fold, the crease
pattern on the finished unit, and
how to assemble.

insert

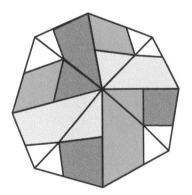

12-unit octahedral
assembly

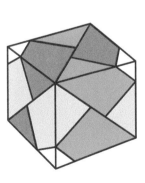

6-unit cube assembly

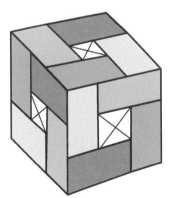

12-unit large cube
assembly

Swan Sonobe

Start by doing Steps 1 through 5 of Daisy Sonobe on page 6.

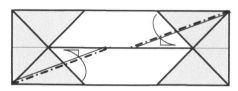

6. Crease and open to bisect angles.

7. Mountain fold existing creases to tuck underneath.

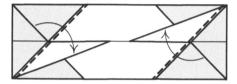

8. Valley fold as shown.

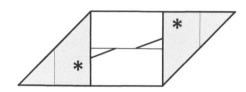

9. Tuck flaps marked ✱ in opening underneath.

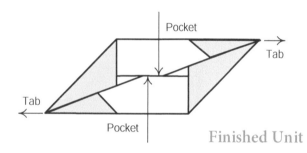

Pocket

Tab

Tab

Pocket

Finished Unit

Refer to pages 2–5 to determine how many units to fold, the crease pattern on the finished unit, and how to assemble.

Assembly

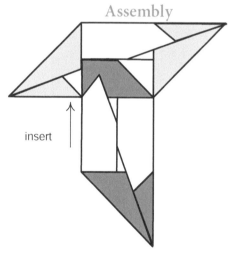

insert

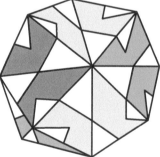

6-unit cube assembly

12-unit octahedral assembly

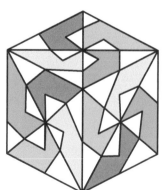

12-unit large cube assembly

Spiked Pentakis Dodecahedron

Make 60 units of any Sonobe variation. This example uses Snow-Capped Sonobe 2 (see page 12). (This 60-unit Sonobe construction was first known to have been made by Michael J. Naughton in the 1980s.)

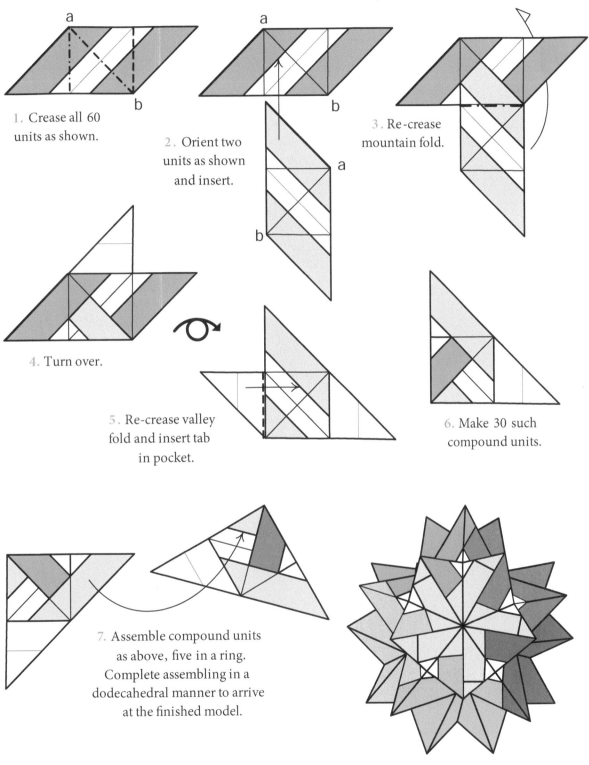

1. Crease all 60 units as shown.

2. Orient two units as shown and insert.

3. Re-crease mountain fold.

4. Turn over.

5. Re-crease valley fold and insert tab in pocket.

6. Make 30 such compound units.

7. Assemble compound units as above, five in a ring. Complete assembling in a dodecahedral manner to arrive at the finished model.

Cosmos Ball Variation (top left), Calla Lily Ball (top right), Phlox Ball (middle), Fantastic (bottom left), and Stella (bottom right).

2 ◆ Enhanced Sonobes

The units made in this chapter are Sonobe-type units but with some enhancements. Hence, they will be called Enhanced Sonobe or eSonobe units. To assemble these units, follow the same general instructions as previously presented in Chapter 1, "Sonobe Variations" (page 1). The Phlox Ball, unlike the other models, will involve additional steps after completion of assembly. The 12-unit and the 30-unit assemblies are recommended for these units. Larger assemblies may not produce pleasing results.

In this chapter we will use rectangles as our starting paper instead of squares. This is a deviation from traditional origami—some might call it evolution. As in so many art forms, deviations often lead to a host of additional possibilities. It is not at all uncommon for origami artist to begin their folding with a rectangle, often with spectacular results. Sometimes I see no need to be a slave to tradition. It is a positive outcome when arts can adapt and change with the influences of new generations of thinkers.

There are origami ways to achieve the rectangles of desired aspect ratios, starting with a square paper. You may refer to the section "Rectangles from Squares" in the Appendix on page 65. When you size one rectangle, use that as a template to cut all the other rectangles needed for your model.

The paper used for these models should not be thicker than the kami variety (regular origami paper available in most craft stores). Paper too thick will lead to uncontrollable holes at the five-point vertices and will diminish the visual impact of the final piece.

Recommendations

Paper Size: Rectangles 3.5"–4.5" in width, length will vary proportionately with model.

Paper Type: Kami or slightly thicker (but not much thicker). Try harmony paper for Cosmos and Calla Lilly Balls.

Finished Model Size

Paper 4" wide yields model of height 4.75".

Cosmos Ball

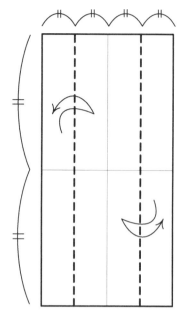

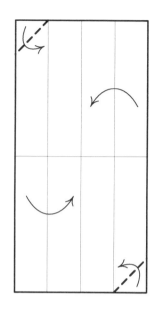

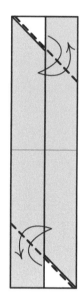

1. Start with 1:2 paper. Crease and open centerlines. Then cupboard fold and open.

2. Fold corners, then re-crease folds from Step 1.

3. Fold and open as shown.

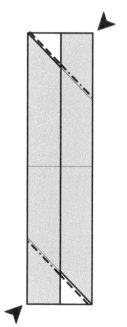

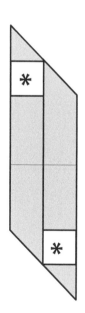

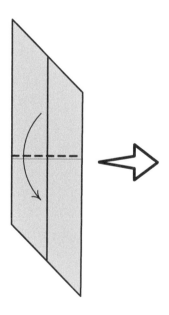

4. Inside reverse fold corners.

5. Tuck areas marked ✳ under flap below.

6. Valley fold pre-existing crease.

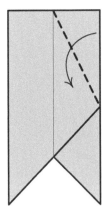

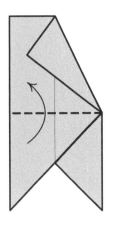

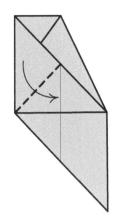

7. Fold corner.

8. Fold flap up.

9. Fold flap down.

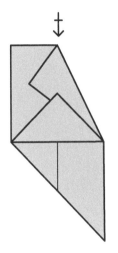

10. Repeat Steps 7–9 on the reverse side.

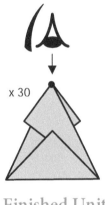

x 30

Finished Unit

Reorient the unit so that you are looking down at the point shown. Gently pull flaps out so that the unit now looks like the figure below.

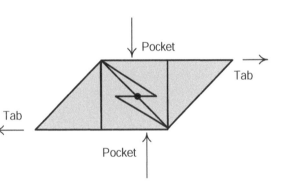

Pocket

Tab

Tab

Pocket

Assemble the 30 units as explained on page 4.

Cosmos Ball Variation

Start with a Cosmos Ball finished unit as described on the previous page.

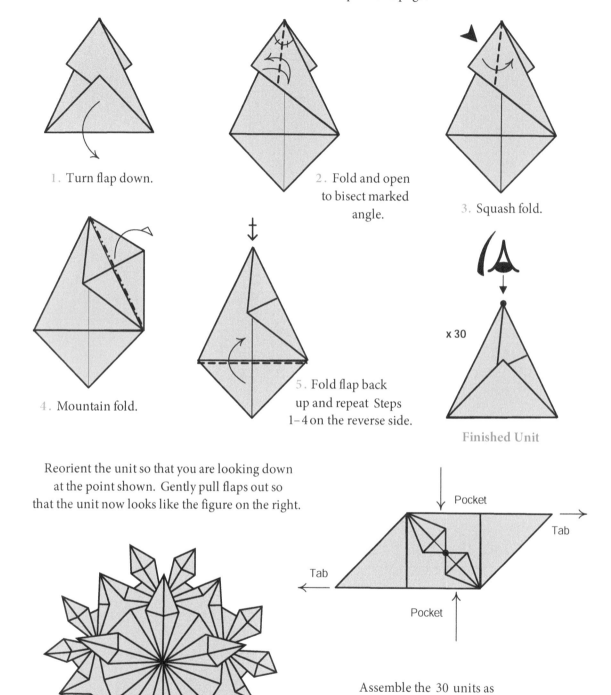

1. Turn flap down.

2. Fold and open to bisect marked angle.

3. Squash fold.

4. Mountain fold.

5. Fold flap back up and repeat Steps 1–4 on the reverse side.

x 30

Finished Unit

Reorient the unit so that you are looking down at the point shown. Gently pull flaps out so that the unit now looks like the figure on the right.

Pocket

Tab

Tab

Pocket

Assemble the 30 units as explained on page 4.

Calla Lily Ball

Start with a regular Cosmos Ball unit as described on page 18 (do all Steps 1 through 10).

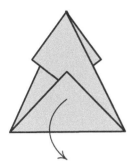

1. Turn flap down.

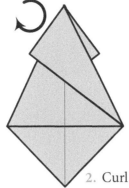

2. Curl upper flap tightly into a cone, inwards.

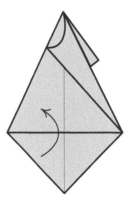

3. Lift flap back up.

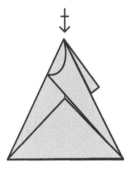

4. Repeat Steps 1–3 on the reverse side.

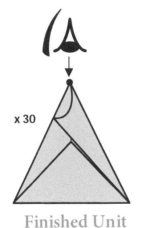

x 30

Finished Unit

Reorient the unit so that you are looking down at the point shown. Gently pull flaps out so that the unit now looks like the figure below.

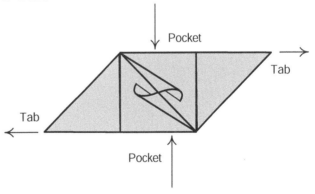

Pocket

Tab

Tab

Pocket

Assemble the 30 units as explained on page 4.

Phlox Ball

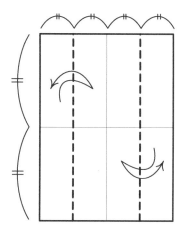

1. Start with 3:4 paper. Crease and open centerlines. Then cupboard fold and open.

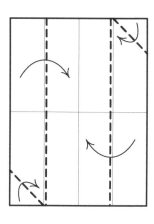

2. Fold corners and re-crease cupboard folds.

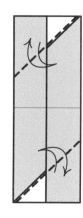

3. Fold and open as shown.

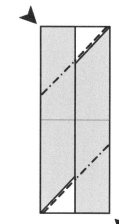

4. Inside reverse fold corners.

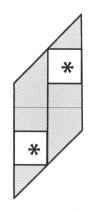

5. Tuck areas marked ✱ under flap below.

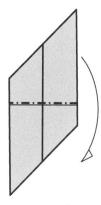

6. Mountain fold pre-existing crease.

7. Crease and open as shown.

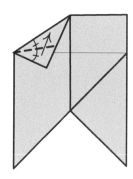

8. Fold corner as shown.

9. Fold to bisect marked angle.

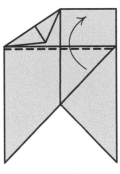

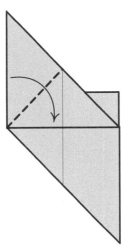

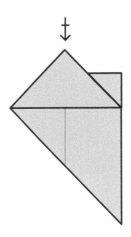

10. Fold flap
back up.

11. Fold as shown.

12. Repeat Steps 7–11
on the reverse side.

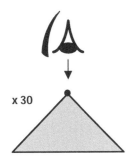

x 30

Finished Unit

Assemble the 30 units as
explained on page 4.

Reorient the unit so that you are looking down at
the point shown. Gently pull flaps out so that
the unit now looks like the figure below. Note that
flaps A and B need to be pulled out AFTER assembly.

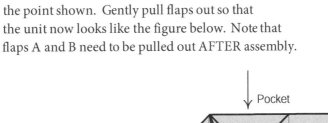

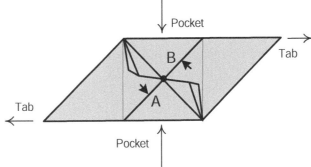

Initial assembly
looks like model on
the left. "Bloom"
petals as described
above to arrive at
the finished model.

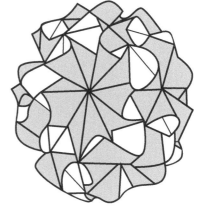

Phlox before blooming

Phlox after blooming

Start with 2:3 paper and do Steps 1 through 7 of the Phlox unit as on page 22.

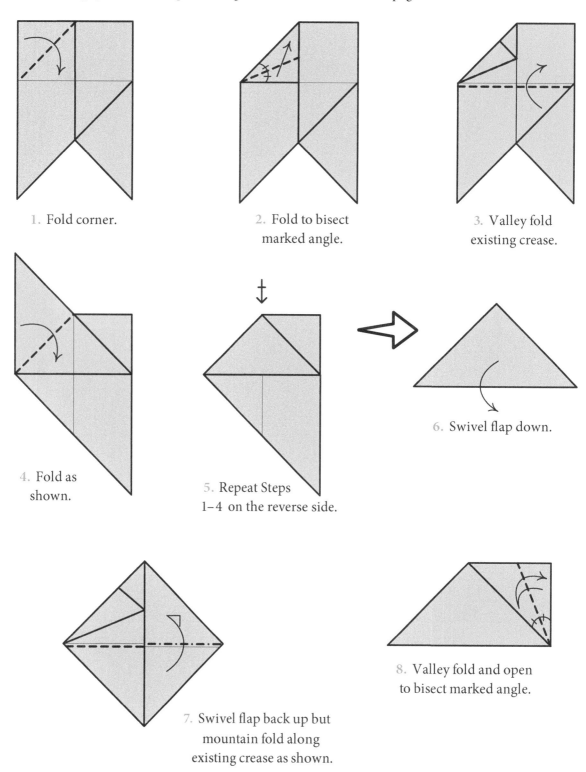

1. Fold corner.

2. Fold to bisect marked angle.

3. Valley fold existing crease.

4. Fold as shown.

5. Repeat Steps 1–4 on the reverse side.

6. Swivel flap down.

7. Swivel flap back up but mountain fold along existing crease as shown.

8. Valley fold and open to bisect marked angle.

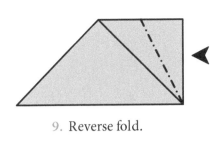

9. Reverse fold.

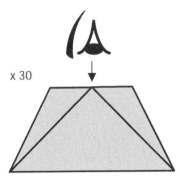

x 30

Finished Unit

10. Repeat Steps 6–9
on the reverse side.

Reorient and assemble as
explained in Phlox Ball.

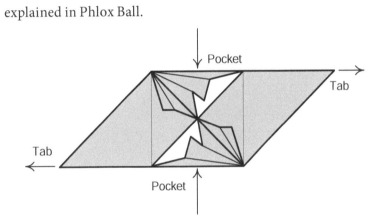

Pocket

Tab

Tab

Pocket

12-unit and 6-unit assemblies of FanTastic.

Start with a finished Phlox unit as on page 22; do all 12 steps.

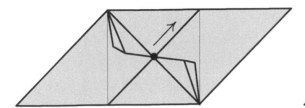

1. Swivel the tip of the unit marked with a dot towards the top right.

2. Valley fold to bisect angles up to where new creases meet. Tip sways back to the top.

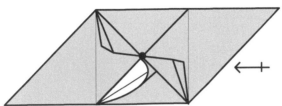

3. Rotate 180° and repeat Steps 1 and 2.

Finished Unit

Pocket

Tab

x30

Tab

Pocket

Assemble like Phlox Ball.

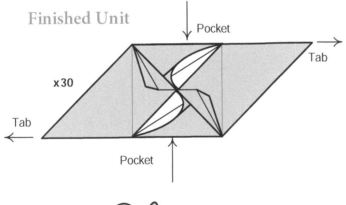

12-unit assembly of Stella

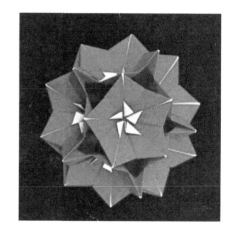

Poinsettia (top left), Plumeria (top right), Passion Flower (middle), Petunia (bottom left), and Primrose (bottom right) Floral Balls.

3 ◆ Floral Balls

My floral balls are inspired by Miyuki Kawamura's Sakura (*Origami Tanteidan 4th Convention Book*, 1998) and Toshikazu Kawasaki's Sakuradama (*Origami Dream World*, 2001). Although my units are much simpler and the locking is quite different, the basic idea that two petals belonging to two adjacent flowers may be formed with one unit, and that 30 of those units may be assembled in a dodecahedral manner to create a floral ball, came from the above mentioned models.

You will notice that all of the floral ball models start with a cupboard fold. The different petal shapes are achieved by altering the folds at the tips of the petals. The center of the flowers can be of two types: exposed lock type and hidden lock type. The exposed lock gives the look and feel of the pistil and stamen of a flower, rendering a much more realistic look. In summary, we will look at five kinds of petal tips and two kinds of centers formed by exposed or hidden locks. Some variation can also be achieved by changing the aspect ratio of the starting paper: the longer the paper, the longer the petals. So, other than those flowers diagrammed here, please go ahead and create new flowers on your own with different combinations that I have not tried out for myself.

Recommendations

Paper Size: Rectangles 3.5"–5" in length. Width will vary proportionately with model.

Paper Type: Everything works except highly glossy paper (need a little friction to hold). Foil paper produces the best locks.

Finished Model Size

Paper 4.5" long yields model of height 5.5".

Poinsettia Floral Ball

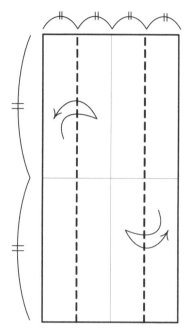

1. Start with 1:2 paper. Fold and open centerlines. Then cupboard fold and open.

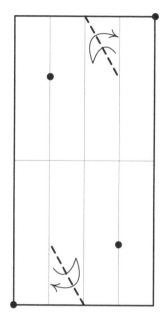

2. Match dots to form new creases as shown.

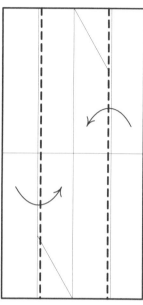

3. Re-crease cupboard folds.

4. Fold and open along existing creases behind.

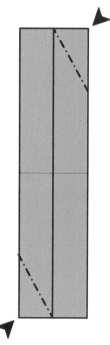

5. Inside reverse fold corners.

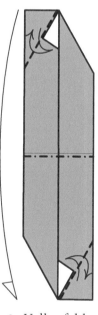

6. Valley fold and open as shown, then mountain fold pre-existing crease.

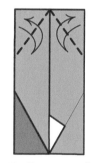

7. Crease and open corners.

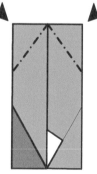

8. Inside reverse fold corners.

Finished Unit

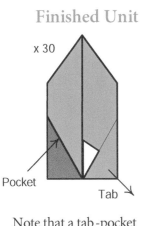

x 30

Pocket

Tab

Note that a tab-pocket
pair is at the back.

Locking

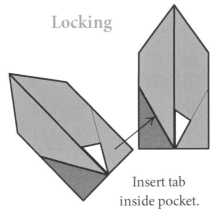

Insert tab
inside pocket.

Firmly crease the two units
together towards the left as
shown in the enlarged inset,
thus locking the units.

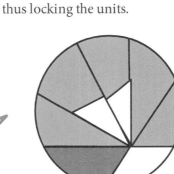

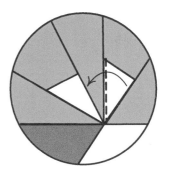

Assembly

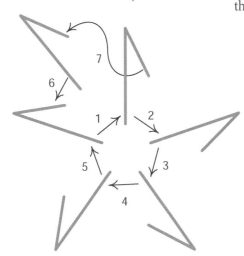

Let each bent line represent one
unit. Assemble 5 units in a circle
to form one flower, following the
sequence numbers. Then insert
a sixth unit doing Steps 6 and 7
to form one hole. Continue
building 12 flowers and 20 holes
to arrive at the finished model.

Passion Flower Ball

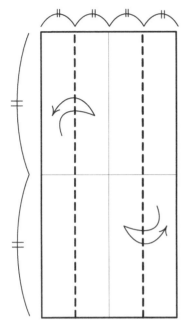

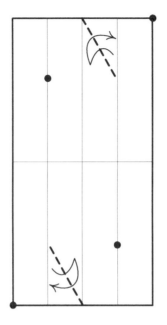

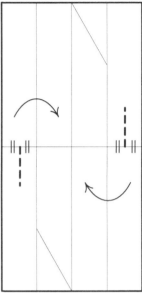

1. Start with 1:2 paper. Fold and open centerlines. Then cupboard fold and open.

2. Match dots to form new creases as shown.

3. Pinch halfway points as shown, then re-crease cupboard folds.

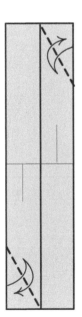

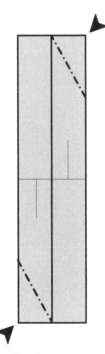

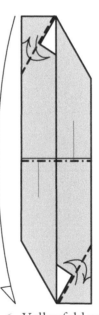

7. Crease and open to match dots.

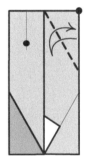

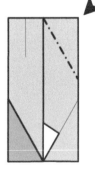

4. Fold and open along existing creases behind.

5. Inside reverse fold corners.

6. Valley fold and open as shown, then mountain fold pre-existing crease.

8. Inside reverse fold corner.

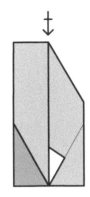

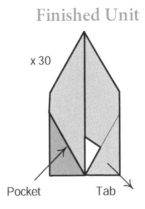

Finished Unit

x 30

Pocket   Tab

Note that a tab-pocket pair is at the back.

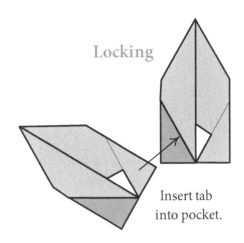

Locking

Insert tab into pocket.

9. Repeat Steps 7 and 8 on the reverse side. The last two reverse folds will overlap.

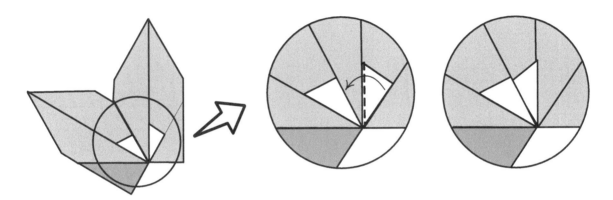

Firmly crease the two units together towards the left as shown in the enlarged inset, thus locking the units.

Assemble as explained in Poinsettia Floral Ball to arrive at the finished model.

Plumeria Floral Ball

Use any rectangle with aspect ratio between silver and bronze rectangles (see the section "Rectangles From Squares" in the Appendix on page 65). This diagram uses 2:3 paper.

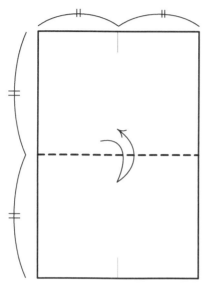

1. Start with 2:3 paper. Fold and open equator. Then pinch top and bottom of center fold.

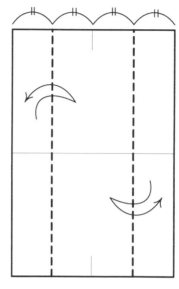

2. Cupboard fold and open.

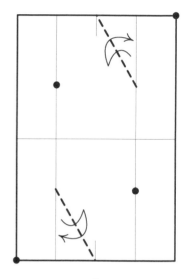

3. Match dots to form new creases.

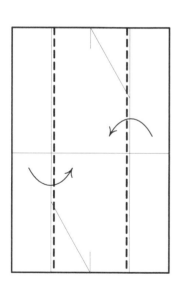

4. Re-crease cupboard folds.

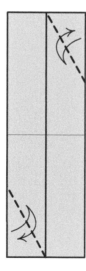

5. Fold and open along existing creases behind.

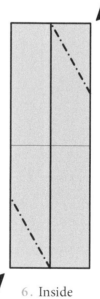

6. Inside reverse fold corners.

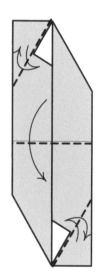

7. Fold and open corners, then valley fold pre-existing crease.

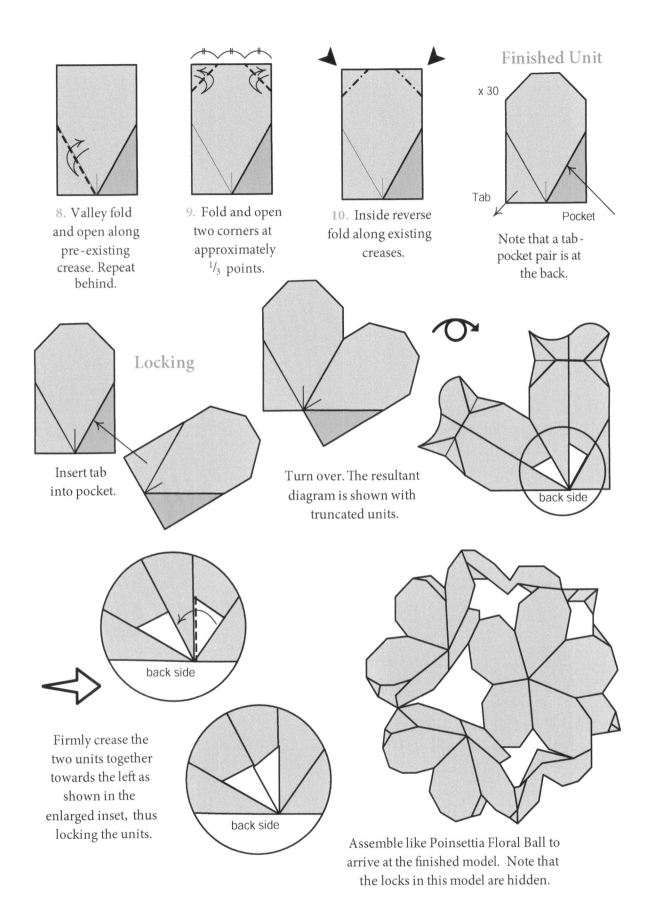

8. Valley fold and open along pre-existing crease. Repeat behind.

9. Fold and open two corners at approximately $1/3$ points.

10. Inside reverse fold along existing creases.

Finished Unit

x 30

Tab

Pocket

Note that a tab-pocket pair is at the back.

Locking

Insert tab into pocket.

Turn over. The resultant diagram is shown with truncated units.

back side

Firmly crease the two units together towards the left as shown in the enlarged inset, thus locking the units.

back side

back side

Assemble like Poinsettia Floral Ball to arrive at the finished model. Note that the locks in this model are hidden.

Petunia Floral Ball

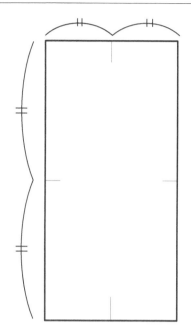

1. Start with 1:2 paper. Pinch ends of center fold and equator, then do Steps 2–6 of Plumeria Floral Ball.

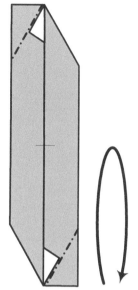

2. Mountain fold and open corners, then curve unit gently towards you, bringing top edge to the bottom. Make curve firm but DO NOT form any crease.

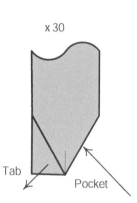

x 30

Tab   Pocket

Finished Unit
(shown truncated)
Other tab-pocket pair is not shown.

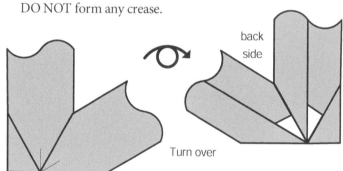

Release curve and insert tab into pocket. Curves will reappear during assembly.

back side

Turn over

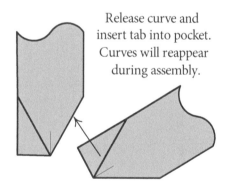

Lock and assemble exactly as in Plumeria Floral Ball (with hidden locks). There will be tension in the units so you may use aid such as miniature clothespins which may be removed upon completion.

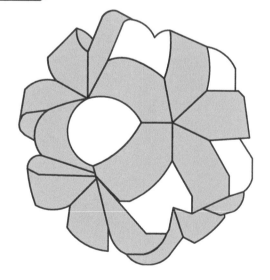

Primrose Floral Ball

Start with 3:4 paper
and fold exactly as
Poinsettia Floral Ball.

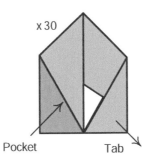

Finished Unit

x 30

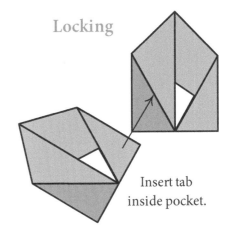

Locking

Pocket   Tab

Note that a tab-pocket
pair is at the back.

Insert tab
inside pocket.

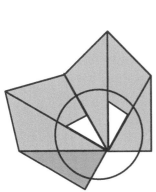

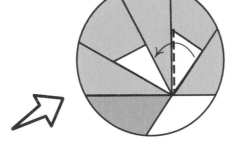

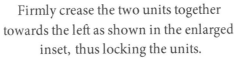

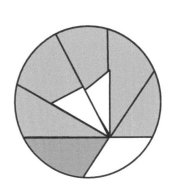

Firmly crease the two units together
towards the left as shown in the enlarged
inset, thus locking the units.

Assemble as explained in
Poinsettia Floral Ball to
arrive at the finished model.

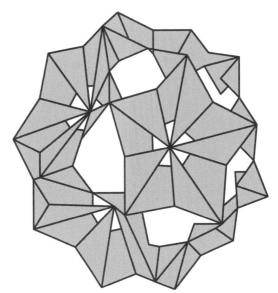

Daisy Dodecahedron 1 (top), 2 (middle), and 3 (bottom).

4 ◆ Patterned Dodecahedra I

In this chapter and the next, we will make various patterned dodecahedra. I first stumbled upon an origami patterned dodecahedron by chance in a Japanese kit that consisted of 30 sheets of paper and directions, all in Japanese—but that is the beauty of origami, it is an international language! The only problem was that I could not read who the creator was. Sometime later I came across some beautiful patterned dodecahedra by Tomoko Fuse. Then, I found other beautiful ones by Silvana Betti Mamino of Italy. Thanks to them all for providing me with the inspiration to create some patterned dodecahedra of my own.

All models in the dodecahedra chapters are made up of 30 units with one unit contributing to two adjacent faces of the dodecahedron. In Chapter 5 ("Patterned Dodecahedra II," page 45), we will incorporate a tiny bit of cutting to create locks as well as use a template to make some final creases on our units.

Recommendations

Paper Size: 3"–5" squares.

Paper Type: Everything works.

Finished Model Size

4" squares yield a model of height 4.75".

Daisy Dodecahedron 1

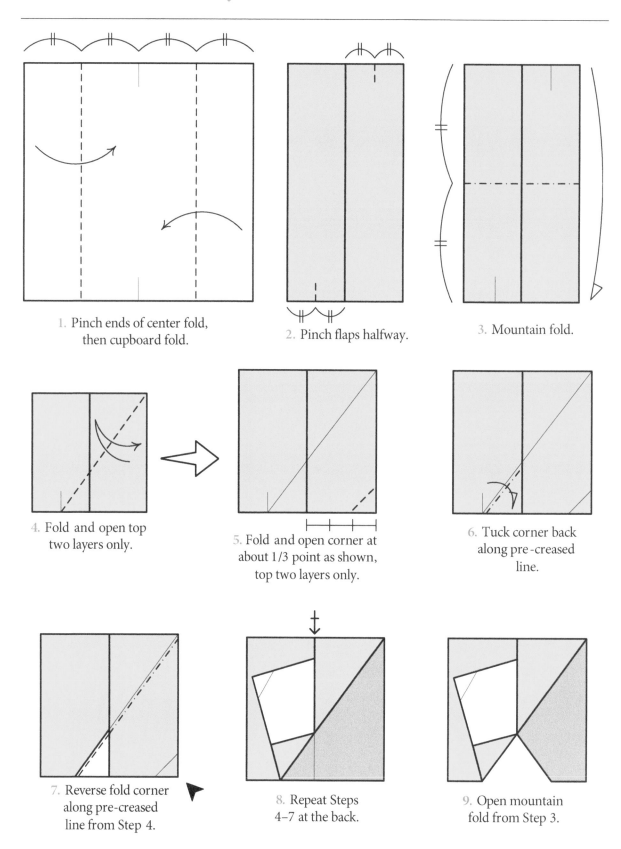

1. Pinch ends of center fold, then cupboard fold.

2. Pinch flaps halfway.

3. Mountain fold.

4. Fold and open top two layers only.

5. Fold and open corner at about 1/3 point as shown, top two layers only.

6. Tuck corner back along pre-creased line.

7. Reverse fold corner along pre-creased line from Step 4.

8. Repeat Steps 4–7 at the back.

9. Open mountain fold from Step 3.

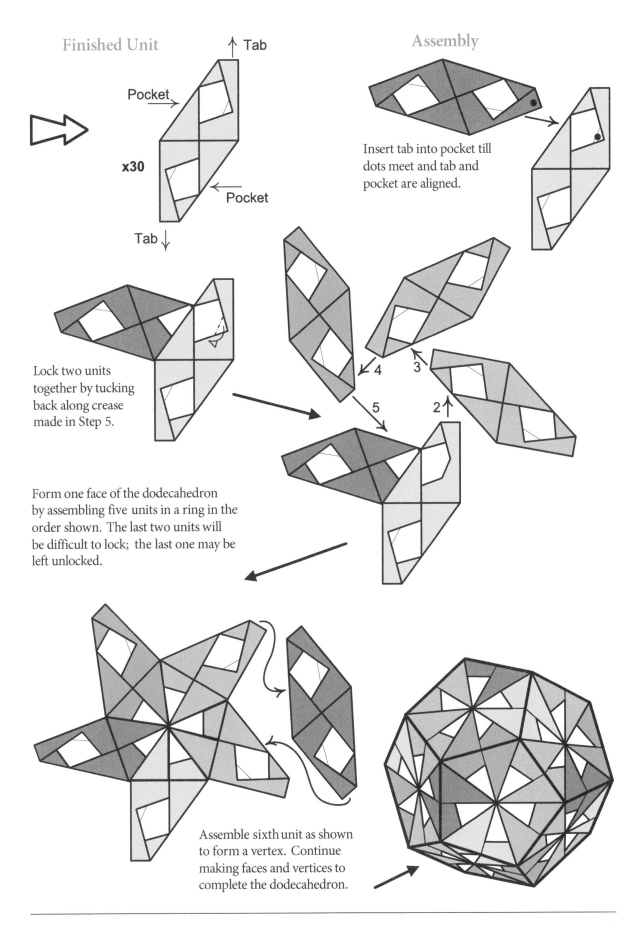

Finished Unit

↑ Tab

Pocket →

x30

← Pocket

Tab ↓

Assembly

Insert tab into pocket till dots meet and tab and pocket are aligned.

Lock two units together by tucking back along crease made in Step 5.

4
3
5
2 ↑

Form one face of the dodecahedron by assembling five units in a ring in the order shown. The last two units will be difficult to lock; the last one may be left unlocked.

Assemble sixth unit as shown to form a vertex. Continue making faces and vertices to complete the dodecahedron.

Daisy Dodecahedron 2

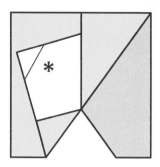

1. Start with Step 9 of the previous model and tuck ✱ marked flap under.

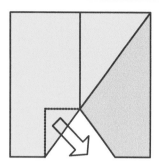

2. Pull corner out from behind.

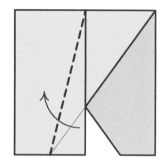

3. Valley fold as shown.

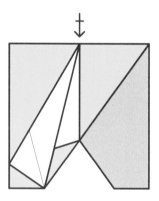

4. Turn over and repeat Steps 1–3.

Finished Unit

x30

Tab

Pocket

Pocket

Tab

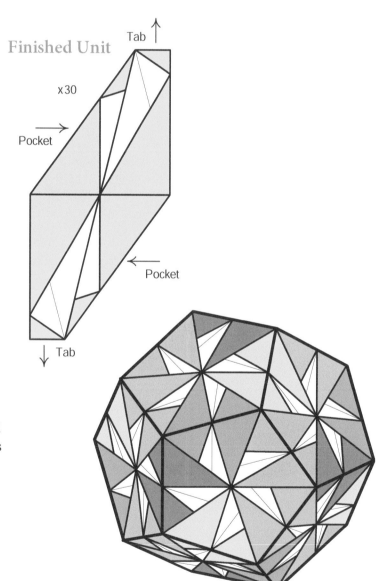

Assemble units as in previous model to arrive at the finished model. Note that locking gets difficult in this model.

Daisy Dodecahedron 3

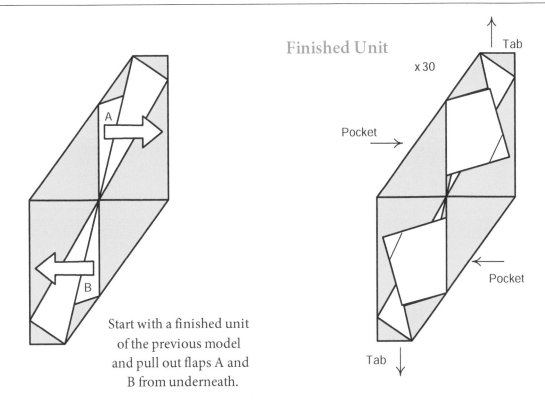

Finished Unit

x 30

Tab

Pocket

Pocket

Tab

Start with a finished unit
of the previous model
and pull out flaps A and
B from underneath.

Assemble units as in previous
dodecahedron to arrive at the
finished model below.

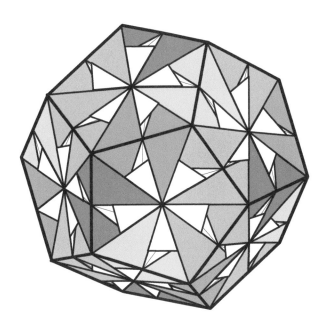

Umbrella (top left), Whirl (top right), Jasmine 2 (middle), and Swirl 1 (bottom left) and 2 (bottom right) Dodecahedra.

5 ◈ Patterned Dodecahedra II

The models in this chapter utilize a template. Use paper of the same size for the template and the units. Four-inch squares will yield finished models about 4" in height. The template will also be created using origami methods. The use of a template will not only expedite the folding process but it will also reduce unwanted creases on a unit, making the finished model look neater.

Making a Template

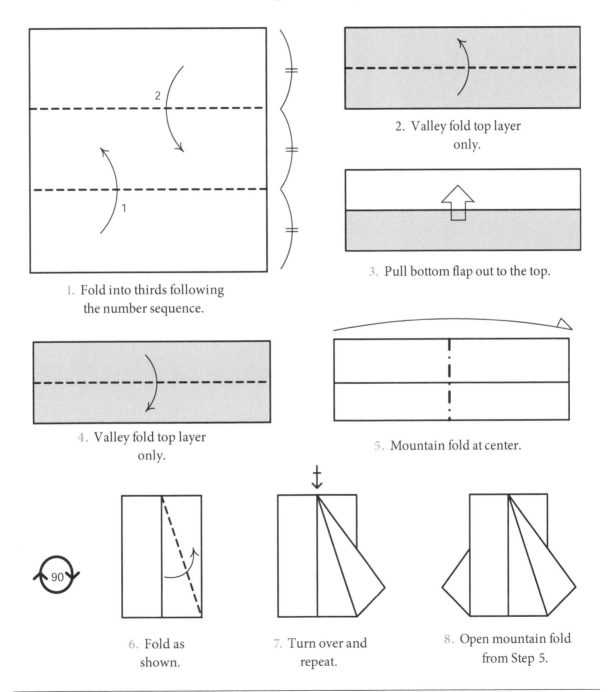

1. Fold into thirds following the number sequence.

2. Valley fold top layer only.

3. Pull bottom flap out to the top.

4. Valley fold top layer only.

5. Mountain fold at center.

6. Fold as shown.

7. Turn over and repeat.

8. Open mountain fold from Step 5.

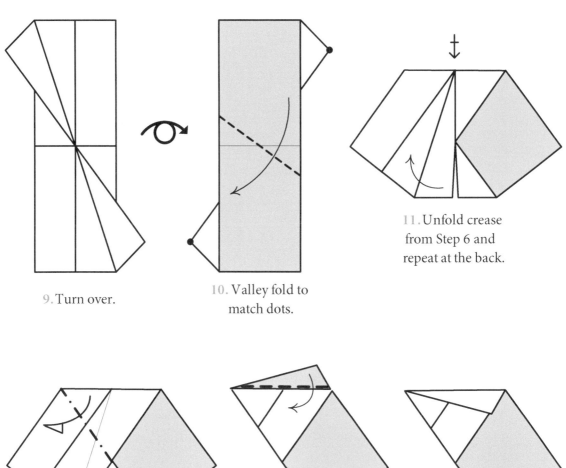

9. Turn over.

10. Valley fold to match dots.

11. Unfold crease from Step 6 and repeat at the back.

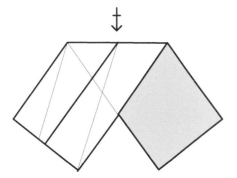

12. Mountain fold along edge of back flap.

13. Valley fold tip.

14. Unfold Steps 13 and 12.

15. Repeat Steps 12–14 at the back and then open to Step 5.

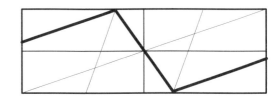

Finished Template Unit
Bold lines represent creases that we are interested in for the actual models.

Umbrella Dodecahedron

First, make one template unit as previously described on page 45. Then, fold 30 units following the diagrams below. Use same size paper for the template and the model.

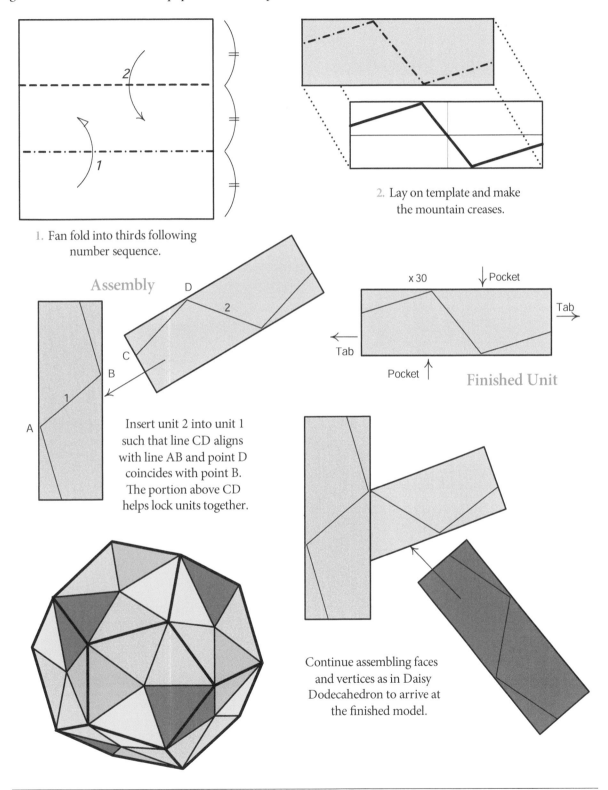

1. Fan fold into thirds following number sequence.

2. Lay on template and make the mountain creases.

Assembly

x 30 Pocket

Tab

Tab

Pocket

Finished Unit

Insert unit 2 into unit 1 such that line CD aligns with line AB and point D coincides with point B. The portion above CD helps lock units together.

Continue assembling faces and vertices as in Daisy Dodecahedron to arrive at the finished model.

Whirl Dodecahedron

First make one template unit as previously described on page 45. Then, fold 30 units following the diagrams below. Use same size paper for the template and the model.

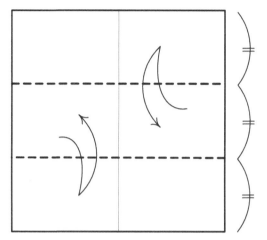

1. Crease and open center fold. Then valley fold into thirds and open.

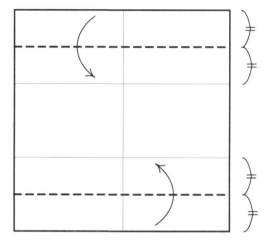

2. Valley fold as shown.

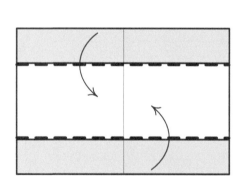

3. Re-crease folds from Step 1.

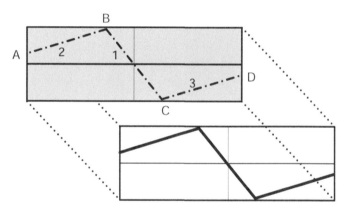

4. Place unit over template and mountain fold and open creases BC, AB, and CD in that order.

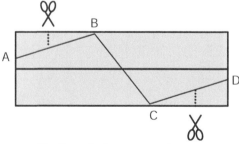

5. To form locks, snip with scissors along the dotted lines, approximately LESS than halfway from A to B. Repeat on DC.

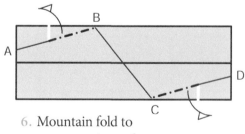

6. Mountain fold to re-crease portions of the flap shown.

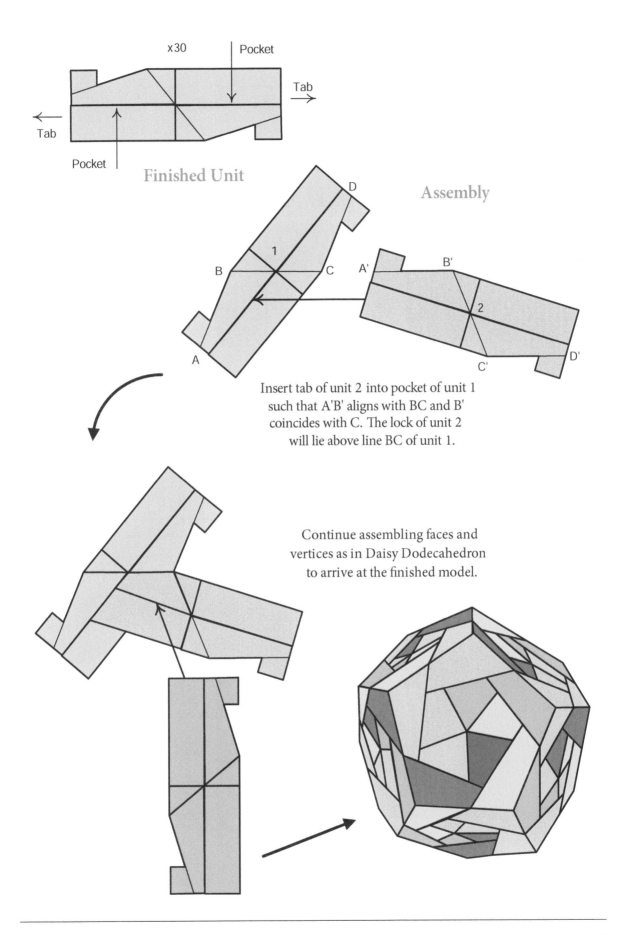

x30

Pocket

Tab

Tab

Pocket

Finished Unit

Assembly

Insert tab of unit 2 into pocket of unit 1
such that A'B' aligns with BC and B'
coincides with C. The lock of unit 2
will lie above line BC of unit 1.

Continue assembling faces and
vertices as in Daisy Dodecahedron
to arrive at the finished model.

Jasmine Dodecahedron 1

First make one template unit as previously described on page 45. Then, fold 30 units following the diagrams below. Use same size paper for the template and the model.

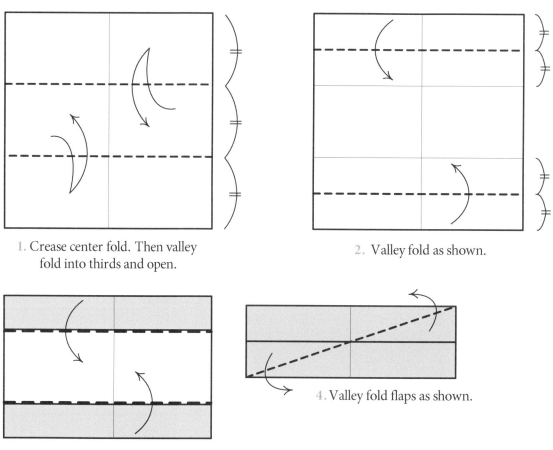

1. Crease center fold. Then valley fold into thirds and open.

2. Valley fold as shown.

3. Re-crease folds from Step 1.

4. Valley fold flaps as shown.

5. Mountain fold as shown.

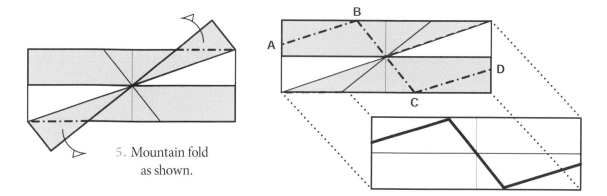

6. Place unit over template and then crease and open mountain folds BC, AB, and CD, in that order.

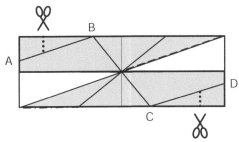

7. Snip with scissors along the dotted lines, approximately less than half way from A to B. Repeat on DC.

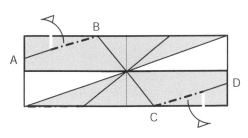

8. Mountain fold to re-crease the portions of AB and CD shown.

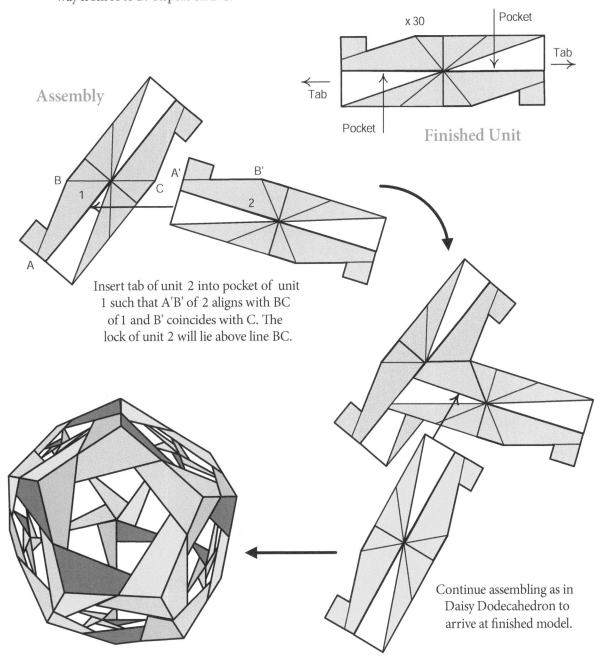

x 30

Pocket

Tab

Tab

Pocket

Finished Unit

Assembly

Insert tab of unit 2 into pocket of unit 1 such that A'B' of 2 aligns with BC of 1 and B' coincides with C. The lock of unit 2 will lie above line BC.

Continue assembling as in Daisy Dodecahedron to arrive at finished model.

Jasmine Dodecahedron 2

First, make one template unit as previously described on page 45. Then, fold 30 units following the diagrams below. Use same size paper for the template and the model.

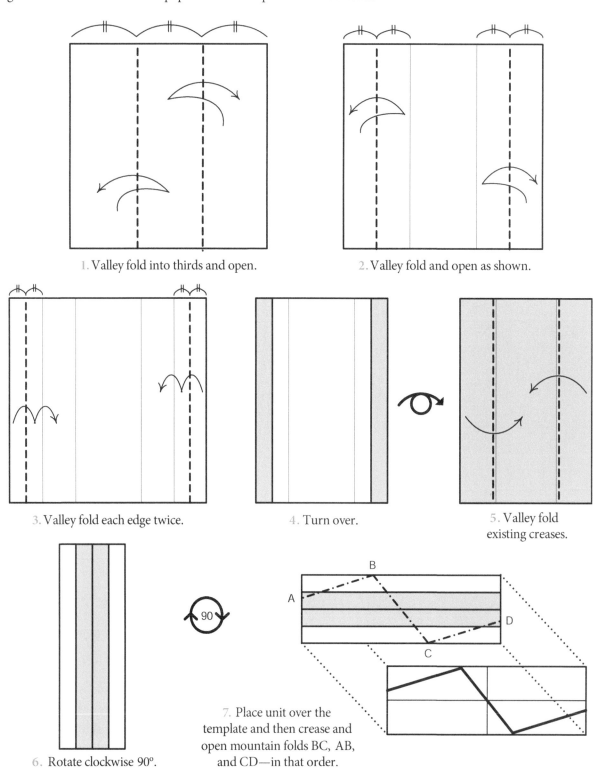

1. Valley fold into thirds and open.

2. Valley fold and open as shown.

3. Valley fold each edge twice.

4. Turn over.

5. Valley fold existing creases.

6. Rotate clockwise 90°.

7. Place unit over the template and then crease and open mountain folds BC, AB, and CD—in that order.

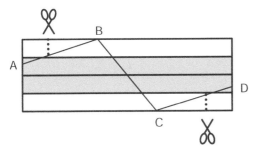

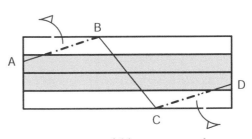

8. Snip with scissors along the dotted lines, approximately less than half way from A to B. Repeat on DC.

9. Mountain fold to re-crease the portions of AB and CD shown.

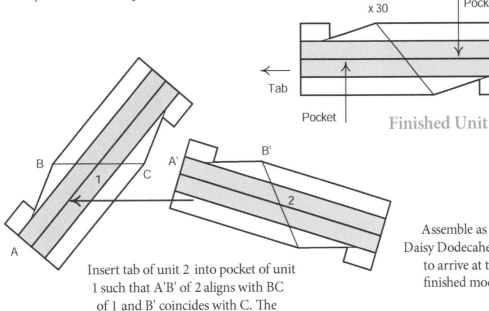

x 30

Pocket

Tab

Tab

Pocket

Finished Unit

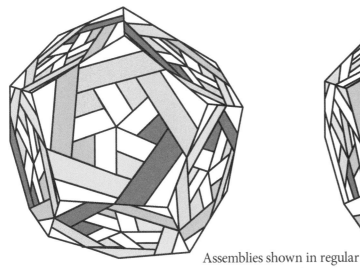

Insert tab of unit 2 into pocket of unit 1 such that A'B' of 2 aligns with BC of 1 and B' coincides with C. The lock of unit 2 will lie above line BC.

Assemble as in Daisy Dodecahedron to arrive at the finished model.

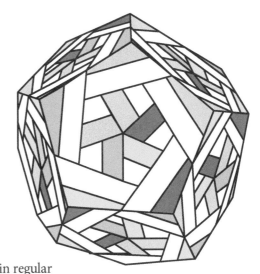

Assemblies shown in regular and reverse colorings.

Jasmine Dodecahedron 3

First, make one template unit as previously described on page 45. Then, fold 30 units following the diagrams below. Use same size paper for the template and the model.

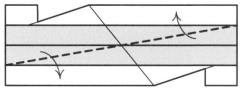

Start with a finished unit of
the previous model and valley
fold the two flaps as shown.

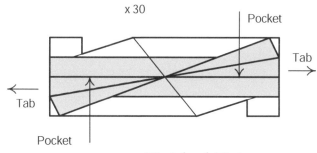

x 30

Pocket

Tab

Tab

Pocket

Finished Unit

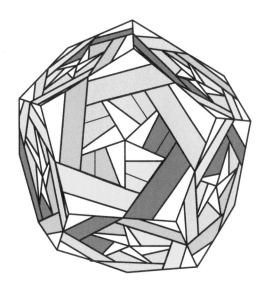

Assemble as in previous
dodecahedron to arrive at
the finished model below.

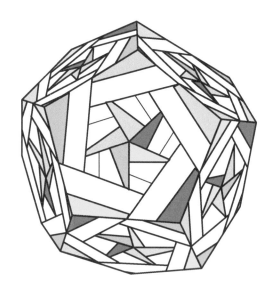

Assemblies shown in regular
and reverse colorings.

Swirl Dodecahedron 1

First, make one template unit as previously described on page 45. Then, fold 30 units following the diagrams below. Use same size paper for the template and the model.

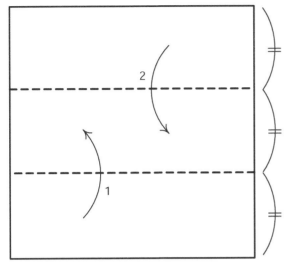

1. Fold into thirds following the number sequence.

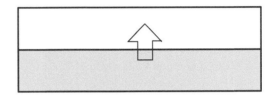

2. Valley fold top layer only.

3. Pull bottom flap out to the top.

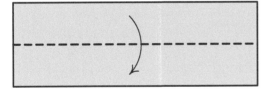

4. Valley fold top layer only.

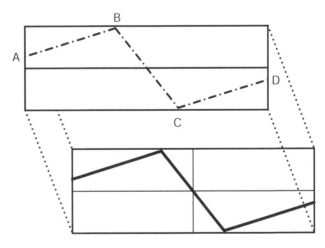

5. Place unit over the template and then crease and open mountain folds BC, AB, and CD—in that order.

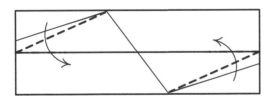

6. Valley fold top layer only.

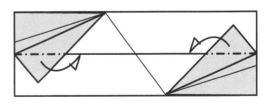

7. Mountain fold corners and tuck under flap.

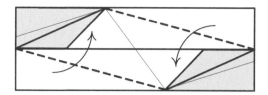

8. Valley fold corners, top layer only.

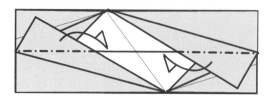

9. Mountain fold corners and tuck under flap.

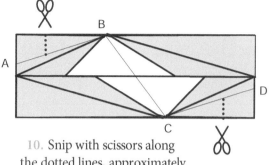

10. Snip with scissors along the dotted lines, approximately less than half way from A to B. Repeat on DC.

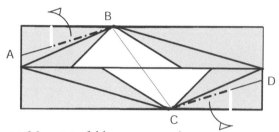

11. Mountain fold to re-crease the portions of AB and CD shown.

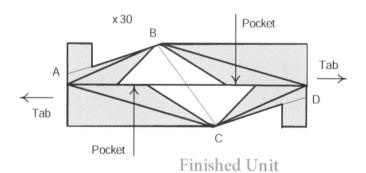

Finished Unit

Assemble as in Whirl Dodecahedron to arrive at the finished model.

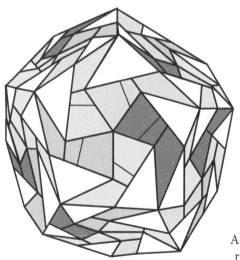

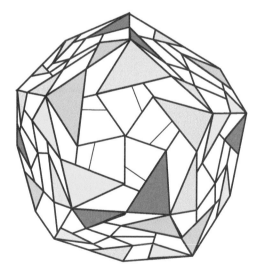

Assemblies shown in regular and reverse colorings.

Swirl Dodecahedron 2

First, make one template unit as previously described on page 45. Then, fold 30 units following the diagrams below. Use same size paper for the template and the model.

Do Steps 1 through 7 of Swirl Dodecahedron 1.

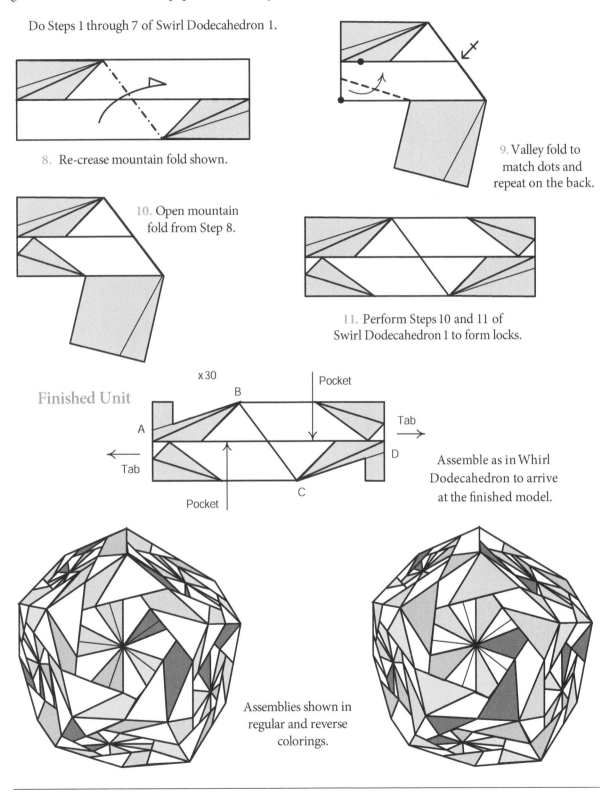

8. Re-crease mountain fold shown.

9. Valley fold to match dots and repeat on the back.

10. Open mountain fold from Step 8.

11. Perform Steps 10 and 11 of Swirl Dodecahedron 1 to form locks.

x 30

Finished Unit

B

Pocket

A

Tab

D

Tab

C

Pocket

Assemble as in Whirl Dodecahedron to arrive at the finished model.

Assemblies shown in regular and reverse colorings.

Lightning Bolt Icosahedral Assembly (top), Star Window (bottom left),
and two views of Twirl Octahedron (bottom middle and right).

6 ◈ Miscellaneous

In this final chapter, three models are presented that do not quite belong to any of the previous chapters. The Lightning Bolt and the Star Windows are essentially Sonobe-type models, but, unlike Sonobe models, they both have openings or windows.

The Twirl Octahedron is reminiscent of Curler Units (2000) by Herman Van Goubergen, but at the time I created this model I was totally unaware of the Curler Units. It was only when someone asked me if I was inspired by the Curler Units that it first came to my attention. Although the models seem similar at first glance, they are really quite different, as you will see. The Curler Units are far more versatile than my Twirl Octahedron units. With the former you can make virtually any polyhedral assembly, but with the latter you can only make an octahedral assembly.

Lightning Bolt

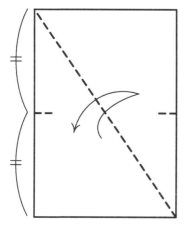

1. Start with A 4 paper. Pinch ends of centerline, then crease and open the diagonal shown.

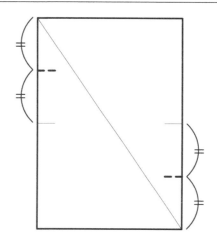

2. Pinch halfway points as marked.

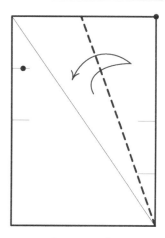

3. Crease and open to match dots.

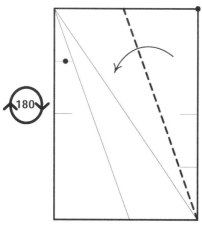

4. Rotate 180° and crease to match dots.

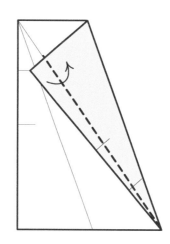

5. Fold to align with diagonal crease from Step 1.

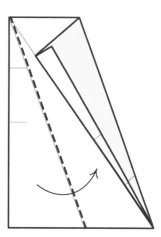

6. Re-crease valley fold and rotate 90°.

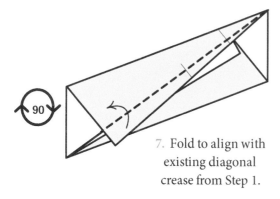

7. Fold to align with existing diagonal crease from Step 1.

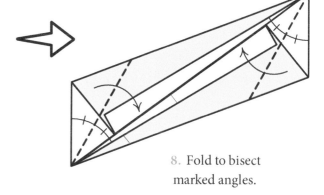

8. Fold to bisect marked angles.

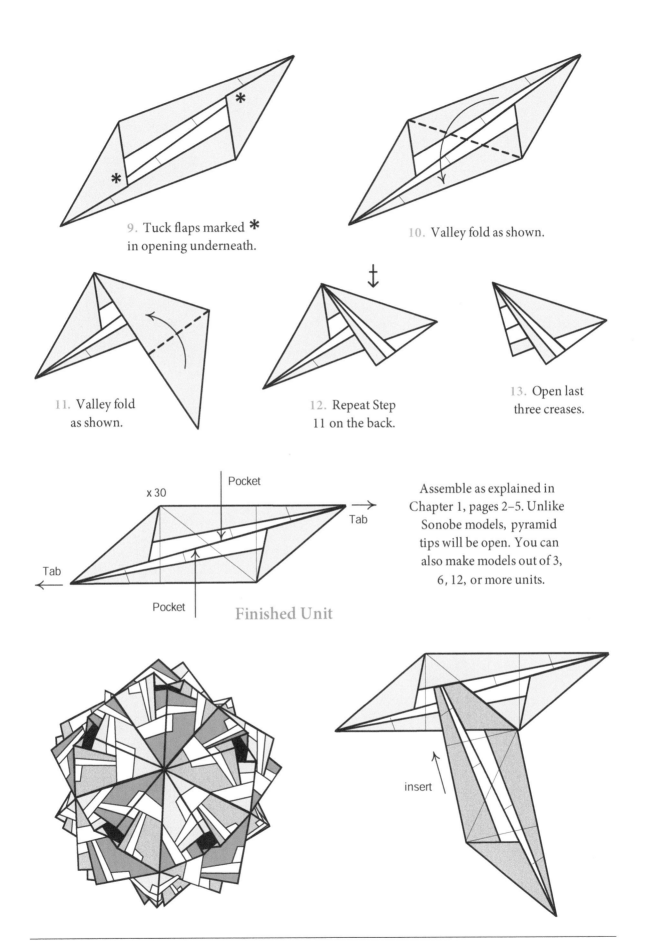

9. Tuck flaps marked ✳ in opening underneath.

10. Valley fold as shown.

11. Valley fold as shown.

12. Repeat Step 11 on the back.

13. Open last three creases.

x 30

Pocket

Tab

Tab

Pocket

Finished Unit

Assemble as explained in Chapter 1, pages 2–5. Unlike Sonobe models, pyramid tips will be open. You can also make models out of 3, 6, 12, or more units.

insert

Twirl Octahedron

This model uses two kinds of units. Use same size paper for both Twirl and Frame Units.

Twirl Units

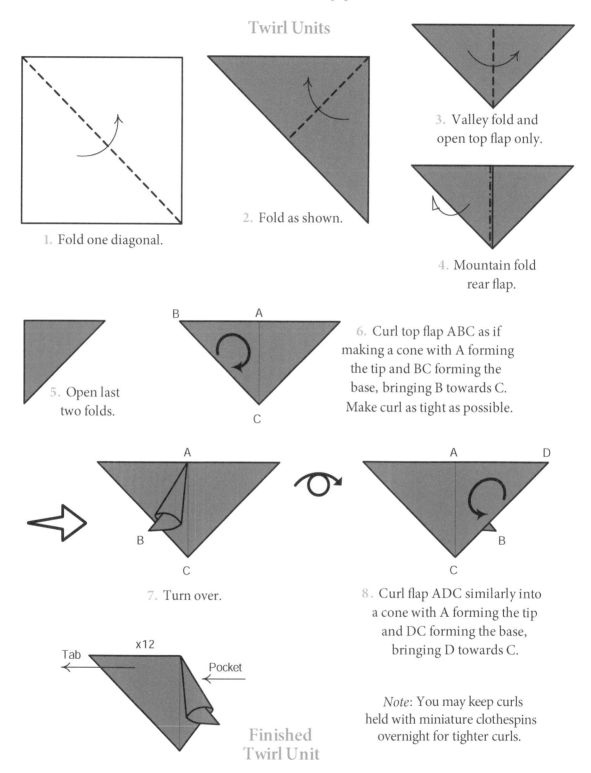

1. Fold one diagonal.

2. Fold as shown.

3. Valley fold and open top flap only.

4. Mountain fold rear flap.

5. Open last two folds.

6. Curl top flap ABC as if making a cone with A forming the tip and BC forming the base, bringing B towards C. Make curl as tight as possible.

7. Turn over.

8. Curl flap ADC similarly into a cone with A forming the tip and DC forming the base, bringing D towards C.

Note: You may keep curls held with miniature clothespins overnight for tighter curls.

x12

Tab

Pocket

Finished Twirl Unit

Frame Units

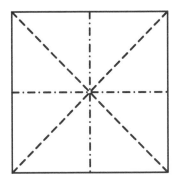

1. Crease and open like waterbomb base.

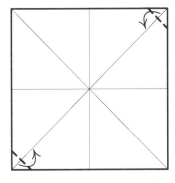

2. Fold small portions of the two corners shown.

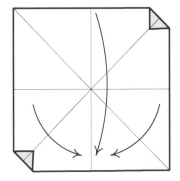

3. Collapse like a waterbomb base.

Finished Frame Unit

x6

Pocket

Tab

Pocket

Tab

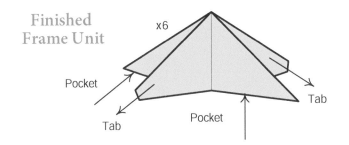

(Note that Step 2 is optional: it is only to distinguish between tabs and pockets.)

Assembly

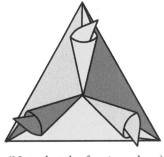

(Note that the face is sunken.)

Making one face of the octahedron.

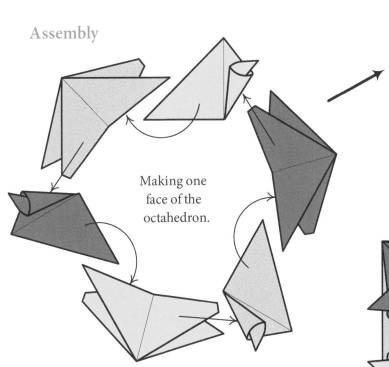

To form one face of the octahedron, connect three Frame Units and three Twirl Units as shown above. Continue forming all eight faces to complete the octahedron.

Star Windows

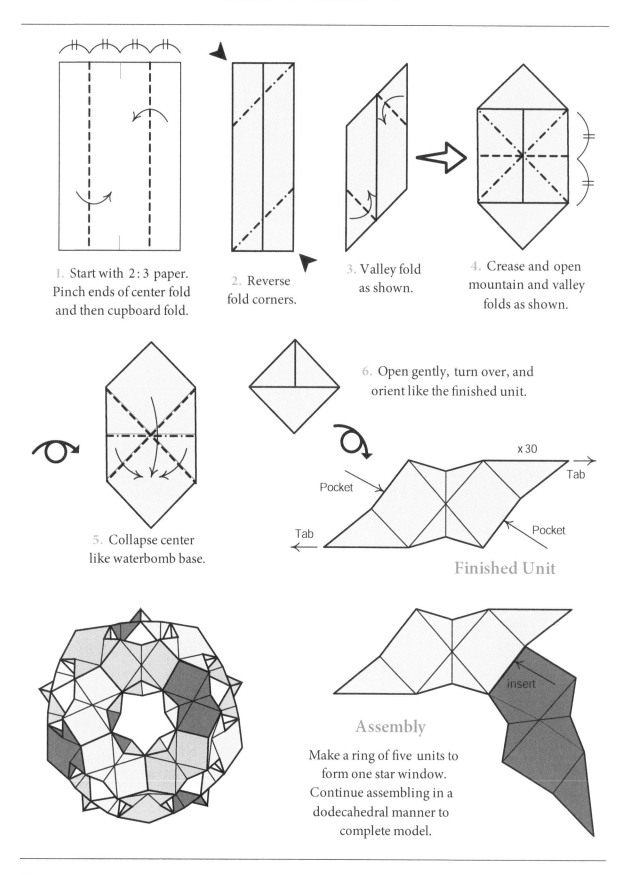

1. Start with 2 : 3 paper. Pinch ends of center fold and then cupboard fold.

2. Reverse fold corners.

3. Valley fold as shown.

4. Crease and open mountain and valley folds as shown.

5. Collapse center like waterbomb base.

6. Open gently, turn over, and orient like the finished unit.

x 30

Tab

Pocket

Tab

Pocket

Finished Unit

Assembly

Make a ring of five units to form one star window. Continue assembling in a dodecahedral manner to complete model.

insert

Appendix

Rectangles from Squares

Aspect Ratio a of a rectangle is defined as the ratio of its width w to its height h, i.e., $a = w{:}h$. Although traditional origami begins with a square, many modern models use rectangles. We will start with a square and show origami ways of arriving at the various rectangles used in this book. The following is organized in ascending order of aspect ratio. Obtain paper sizes as below, and use them as templates to cut papers for your actual units.

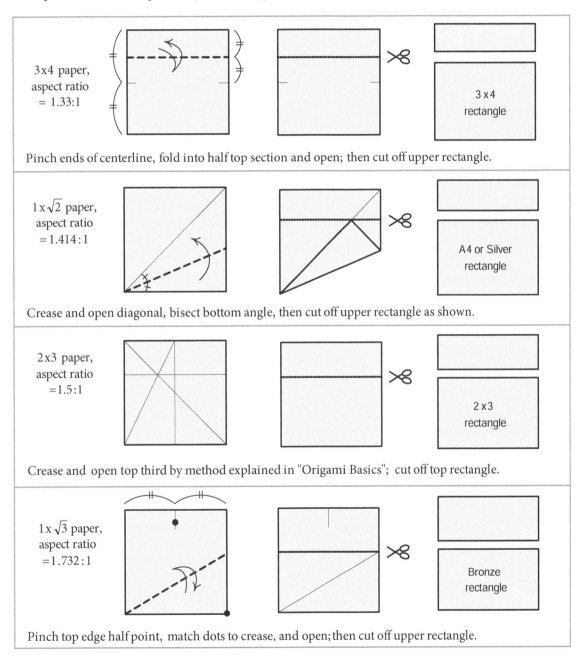

3x4 paper, aspect ratio = 1.33:1

3 x 4 rectangle

Pinch ends of centerline, fold into half top section and open; then cut off upper rectangle.

$1 \times \sqrt{2}$ paper, aspect ratio = 1.414:1

A4 or Silver rectangle

Crease and open diagonal, bisect bottom angle, then cut off upper rectangle as shown.

2x3 paper, aspect ratio =1.5:1

2 x 3 rectangle

Crease and open top third by method explained in "Origami Basics"; cut off top rectangle.

$1 \times \sqrt{3}$ paper, aspect ratio =1.732:1

Bronze rectangle

Pinch top edge half point, match dots to crease, and open; then cut off upper rectangle.

Homogeneous Color Tiling

Homogeneous color tiling is the key to making polyhedral modular origami look additionally attractive. While many models look nice made with a single color, there are many other models that look astounding with the use of multiple colors. Sometimes random coloring works, but symmetry lovers would definitely prefer a homogenous color tiling. Determining a solution such that no two units of the same color are adjacent to each other is quite a pleasantly challenging puzzle. For those who do not have the time or patience, or do not find it pleasantly challenging but love symmetry, I present this section.

In the following figures each edge of a polyhedron represents one unit, dashed edges are invisible from the point of view. It is obvious that for a homogeneous color tiling the number of colors you choose should be a sub-multiple or factor of the number of edges or units in your model. For example, for a 30-unit model, you can use three, five, six or ten colors.

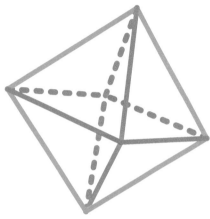

Three-color tiling of an octahedron
(Every face has three distinct colors.)

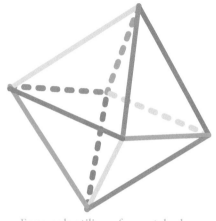

Four-color tiling of an octahedron
(Every vertex has four distinct colors,
and every face has three distinct colors.)

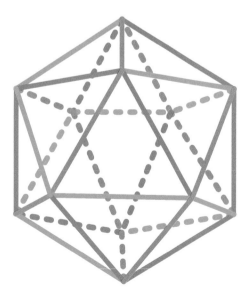

Three-color tiling of an icosahedron
(Every face has three distinct colors.)

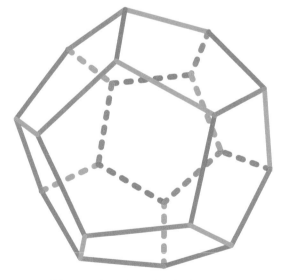

Three-color tiling of a dodecahedron
(Every vertex has three distinct colors.)

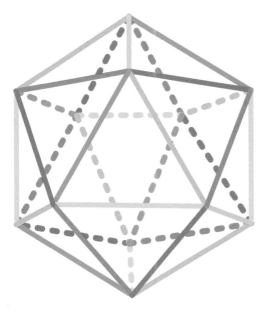

Five-color tiling of an icosahedron
(Every vertex has five distinct colors,
and every face has three distinct colors.)

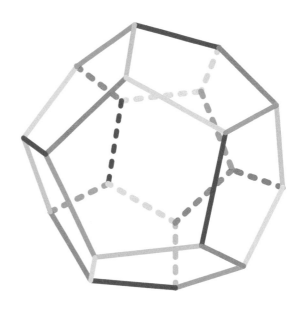

Five-color tiling of a dodecahedron
(Every face has five distinct colors,
and every vertex has three distinct colors.)

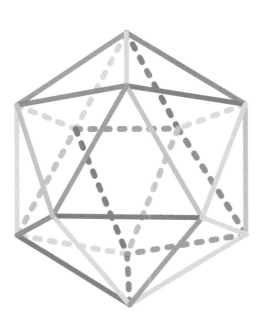

Six-color tiling of an icosahedron
(Every vertex has five distinct colors,
and every face has three distinct colors.)

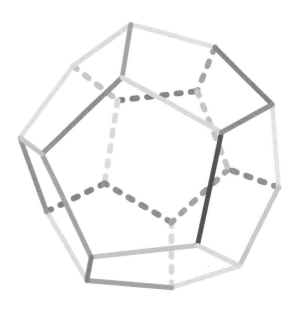

Six-color tiling of a dodecahedron
(Every face has five distinct colors,
and every vertex has three distinct colors.)

Origami, Mathematics, Science, and Technology

Mathematics is everywhere we look, and that includes origami. Although most people conceive origami as child's play and libraries and bookstores generally shelve origami books in the juvenile section, there is a profound connection between mathematics and origami. Hence, it is not surprising that so many mathematicians, scientists, and engineers have shown a keen interest in this field—some have even taken a break from their regular careers to delve deep into the depths of exploring the connection between origami and mathematics. So strong is the bond that a separate international origami conference, *Origami Science, Math, and Education* (OSME) has been dedicated to its cause since 1989. Professor Kazuo Haga of University of Tsukuba, Japan, rightly proposed the term o*rigamics* in 1994 to refer to this genre of origami that is heavily related to science and mathematics.

Background

The mathematics of origami has been extensively studied not only by origami enthusiasts around the world but also by mathematicians, scientists, and engineers, as well as artists. These interests date all the way back to the late nineteenth century when Tandalam Sundara Row of India wrote the book *Geometric Exercises in Paper Folding* in 1893. He used novel methods to teach concepts in Euclidean geometry simply by using scraps of paper and a penknife. The geometric results were so easily attainable that the book became very attractive to teachers as well as students and is still in print to this day. It also inspired quite a few mathematicians to investigate the geometry of paper folding. Around the 1930s, Italian mathematician Margherita Beloch found out that origami can do more than straightedge-and-compass geometric constructions and discovered something similar to one of the Huzita-Hatori Axioms explained next.

While a lot of studies relating origami to mathematics were going on, nobody actually formalized any of them into axioms or theorems until 1989 when Japanese mathematician Humiaki Huzita and Italian mathematician Benedetto Scimemi laid out a list of six axioms to define the algebra and geometry of origami. Later, in 2001, origami enthusiast Hatori Koshiro added a seventh axiom. Physicist, engineer, and leading origami artist Dr. Robert Lang has proved that these seven axioms are complete, i.e., there are no other possible ways of defining a single fold in origami using alignments of points and lines. All geometric origami constructions involving single-fold steps that can be serialized can be performed using some or all of these seven axioms. Together known as the Huzita-Hatori Axioms, they are listed below:

1. Given two lines L_1 and L_2, a line can be folded placing L_1 onto L_2.

2. Given two points P_1 and P_2, a line can be folded placing P_1 onto P_2.

3. Given two points P_1 and P_2, a line can be folded passing through both P_1 and P_2.

4. Given a point P and a line L, a line can be folded passing through P and perpendicular to L.

5. Given two points P_1 and P_2 and a line L, a line can be folded placing P_1 onto L and passing through P_2.

6. Given two points P_1 and P_2 and two lines L_1 and L_2, a line can be folded placing P_1 onto L_1 and placing P_2 onto L_2.

7. Given a point P and two lines L_1 and L_2, a line can be folded placing P onto L_1 and perpendicular to L_2.

Soon after, mathematician Toshikazu Kawasaki proposed several origami theorems, the most well known one being the Kawasaki Theorem, which states that if the angles surrounding a single vertex in a flat origami crease pattern are a_1, a_2, a_3, ..., a_{2n}, then:

$$a_1 + a_3 + a_5 + ... + a_{2n-1} = 180°$$

and

$$a_2 + a_4 + a_6 + ... + a_{2n} = 180°$$

Stated simply, if one adds up the values of every alternate angle formed by creases passing through a point, the sum will be 180°. This can be demonstrated by geometry students playing around with some paper. Discoveries similar to the Kawasaki Theorem have also been made independently by mathematician Jacques Justin.

Physicist Jun Maekawa discovered another fundamental origami theorem. It states that the difference between the number of mountain creases (M) and the number of valley creases (V) surrounding a vertex in the crease pattern of a flat origami is always 2. Stated mathematically, $|M − V| = 2$. There are some more fundamental laws and theorems of mathematical origami, including the laws of layer ordering by Jacques Justin, which we will not enumerated here. Other notable contributions to the foundations of mathematical origami have been made by Thomas Hull, Shuji Fujimoto, K. Husimi, M. Husimi, sarah-marie belcastro, Robert Geretschläger, and Roger Alperin.

The use of computational mathematics in the design of origami models has been in practice for many years now. Some of the first known algorithms for such designs were developed by computer scientists Ron Resch and David Huffman in the sixties and seventies. In more recent years Jun Maekawa, Toshiyuki Meguro, Fumiaki Kawahata, and Robert Lang have done extensive work in this area and took it to a new level by developing computer algorithms to design very complex origami models. Particularly notable is Dr. Lang's well-known TreeMaker algorithm, which produces results that are surprisingly lifelike.

Due to the fact that paper is an inexpensive and readily available resource, origami has become extremely useful for the purposes of modeling and experimenting. There are a large number of educators who are using origami to teach various concepts in a classroom. Origami is already being used to model polyhedra in mathematics classes, viruses in biology classes, molecular structures in chemistry classes, geodesic domes in architecture classes, DNA in genetics classes, and crystals in crystallography classes. Origami has its applications in technology as well, some examples being automobile airbag folding, solar panel folding in space satellites, design of some optical systems, and even gigantic foldable telescopes for use in the geostationary orbit of the Earth. Professor Koryo Miura and Dr. Lang have made major contributions in these areas.

Modular Origami and Mathematics

As explained in the preface, modular origami almost always means polyhedral or geometric modular origami, although there are some modulars that have nothing to do with geometry or polyhedra. For the purposes of this essay, we will assume that *modular origami* refers to *polyhedral modular origami*. A polyhedron is a three-dimensional solid that is bound by polygonal faces. A polygon in turn is a two-dimensional figure bound by straight lines. Aside from the real polyhedra themselves, modular origami involves the construction of a host of other objects that are based on polyhedra.

Looking at the external artistic, often floral appearance of modular origami models, it is hard to believe that there is any mathematics involved at all. But, in fact, for every model there is an underlying polyhedron. It could be based on the Platonic or Archimedean solids (see page xv), some of which have been with us since the BC period, or it could be based on prisms, antiprisms, Kepler-Poinsot solids, Johnson's Solids, rhombohedra, or even irregular polyhedra. The most referenced polyhedra for origami constructions are undoubtedly the five Platonic solids.

Assembly of the units that comprise a model may seem very puzzling to the novice, but understanding certain mathematical aspects can considerably simplify the process. First, one must determine whether a unit is a face unit, an edge unit or a vertex unit, i.e., does the unit identify with a face, an edge, or a vertex of the underlying polyhedron. Face units are the easiest to identify. Most modular model units and all models presented in this book except the *Twirl Octahedron* use edge units. There are only a few known vertex units, some examples being David Mitchell's *Electra, Gemini and Proteus*, Ravi Apte's *Universal Vertex Module (UVM)*, the *Curler Units* by Herman Goubergen, and the models presented in the book *Multimodular Origami Polyhedra* by Rona Gurkewitz and Bennett Arnstein. For edge units there is a second step involved—one must identify which part of the unit (which is far from looking like an edge) actually associates with the edge of a polyhedron. Unfortunately, most of the modular origami creators do not bother to specify whether a unit is of face type, edge type, or vertex type because it is generally perceived to be quite intuitive. But, to a beginner, it may not be so intuitive. On closer observation and with some amount of trial and error, though, one may find that it is not so difficult after all. Once the identifications are made and the folder can see through the maze of superficial designs and perceive the unit as a face, an edge, or a vertex, assembly becomes much simpler. Now it is just a matter of following the structure of the polyhedron to put the units in place. With enough practice, even the polyhedron chart need not be consulted anymore.

The Design Process

How flat pieces of paper can be transformed into such aesthetically pleasing three-dimensional models simply by folding is always a thing to ponder. Whether it is a one-piece model or an interlocked system of modules made out of several sheets of paper, it is equally mind-boggling. While designing a model, one might start thinking mathematically and proceed from there. Alternatively, one might create a model intuitively and later open it back out into the flat sheet of paper from which it originated and read into the mathematics of the creases. Further refinement in the design process if required can be achieved after having studied the creases. Either way of designing and creating is used by origami artists.

In this section we will illustrate how some simple geometry was used in the design of some of the units. Chapters 4 and 5 on patterned dodecahedra are perhaps the best examples. The units were first obtained by trial and error and then geometric principles were used to prove that the desired results had been obtained.

We know that a dodecahedron is made up of 12 regular pentagonal faces. We also know from geometry that the sum of all the internal angles of a polygon with n sides is $2n - 4$ right angles. So, for a pentagon it would be $(2 \times 5) - 4$ right angles = 6 right angles = 540°. Hence, each angle of a regular pentagon is 540°/5 = 108 °.

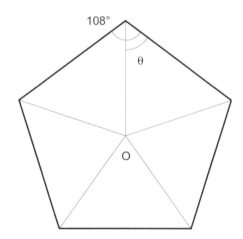

Each face of the patterned dodecahedra presented in this book is sectioned into five equal isosceles triangles as shown above, with O being the center of the pentagon;

∴ θ = 108° / 2 = 54°.

Now consider one unit of a patterned dodecahedron in Chapter 4 as shown below. We will prove that the marked angle θ which translates to θ on the previous page is approximately equal to 54°.

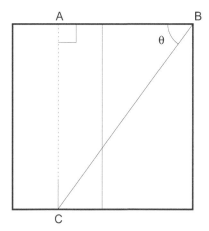

The figure above is from Step 5 of Daisy Dodecahedron 1 (page 40). Let the side of the starting square be of length a.

Draw a perpendicular (⊥) from C to meet the opposite edge at A.

From the folding sequence we know that

$$AC = a/2 \text{ and } AB = a/8 + a/4 = 3a/8.$$

From △ ABC,

$$\tan θ = AC/AB$$

$$= (a/2)/(3a/8) = 4/3;$$

∴ θ = $\tan^{-1}(4/3)$ = 53.1° ≈ 54°.

Next we will consider one template unit of a patterned dodecahedron in Chapter 5 (see pages 45–46).

The figure on the upper right shows a patterned dodecahedron template unit unfolded to Step 9 with not all creases shown. We will prove that the marked angle θ which translates to θ in the sectioned pentagon shown earlier is approximately equal to 54°. Let the side of the starting square be of length a, and let ∠AOC = ϕ.

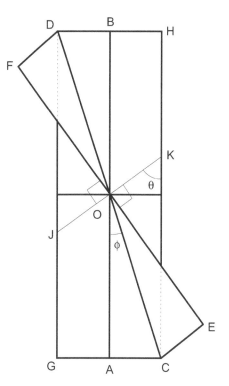

From the folding sequence we know that AO = $a/2$ and AC = $a/6$. We also know that JK ⊥ EF.

From △ AOC, tan ϕ = (a/6)/(a/2) = 1/3;

∴ ϕ = $\tan^{-1}(1/3)$.

△ AOC ≅ △ COE (self explanatory if you look at Step 6 of template folding);

∴ ∠COE = ∠AOC = ϕ.

Since HC ∥ BA and OK intersects both,

∠JOA = ∠OKC = θ (corresponding angles).

Now we know the sum of all angles that make up the straight angle ∠JOK = 180°.

So, ∠JOA + ∠AOC + ∠COE + ∠EOK = 180°

or θ + ϕ + ϕ + 90° = 180°

or θ + 2ϕ = 90°;

∴ θ = 90° – 2ϕ = 90° – 2 x $\tan^{-1}(1/3)$ = 53.1° ≈ 54°.

For both of the examples presented, 53.1° ≈ 54° is a close enough approximation for the purposes of these origami dodecahedra constructions.

1. The figure shows the creases for the traditional method of folding a square sheet of paper into thirds as presented on page xii. ABCD is the starting square, and E and F are midpoints of AB and CD, respectively. The line GH runs through the intersection point of AC and DE parallel to edge AB. Prove that AG is 1/3 of AD.

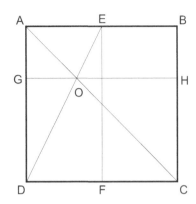

2. The figure shows the creases for the traditional method of obtaining a Silver Rectangle (1 × √2) from a square sheet of paper as presented on page 65. ABCD is the starting square and DG is the bisecting crease of ∠BDC, with O being the shadow of C at the time of creasing. EF runs parallel to AB through point O. Prove that ED is 1/√2 of AD.

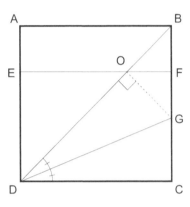

3. The figure shows the creases for the traditional method of obtaining a Bronze Rectangle (1 × √3) from a square sheet of paper as presented on page 65. ABCD is the starting square, and G is the midpoint of AB. Crease DF is obtained by bringing point C to the ⊥ coming down from G. OD and OF are the shadows of CD and CF, respectively, at the time of the creasing. EF runs parallel to AB. Prove that ED is 1/√3 of AD.

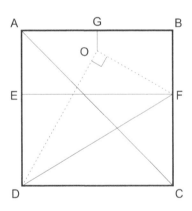

4. The Toshie's Jewel presented in Chapter 1 ("Sonobe Variations") is a triangular dipyramid with each face being a right isosceles triangle. The construction has been made with three Sonobe units. Construct a similar polyhedron with six Sonobe units.

Suggested Reading

Rick Beech, *Origami: The Complete Practical Guide to the Ancient Art of Paperfolding*, Lorenz Books, 2001.

Margherita Beloch, "Sul metodo del ripiegamento della carta per la risoluzione dei problemi geometrici" ("On paper folding methods for the resolution of geometric problems," in Italian), *Periodico di Mathematiche*, Series IV, vol. 16, no. 2, 1936, pp. 104–108.

H. S. M. Coxeter, *Regular Polytopes*, Reprinted by Dover Publications, 1973.

Alexandra Dirk, *Origami Boxes for Gifts, Treasures and Trifles*, Sterling, 1997.

Tomoko Fuse, *Fabulous Origami Boxes*, Japan Publications, 1998.

Tomoko Fuse, *Joyful Origami Boxes*, Japan Publications, 1996.

Tomoko Fuse, *Kusudama Origami*, Japan Publications, 2002.

Tomoko Fuse, *Origami Boxes*, Japan Publications, 1989.

Tomoko Fuse, *Quick and Easy Origami Boxes*, Japan Publications, 2000.

Tomoko Fuse, *Unit Origami: Multidimensional Transforms*, Japan Publications, 1990.

Tomoko Fuse, *Unit Polyhedron Origami*, Japan Publications, 2006.

Rona Gurkewitz and Bennett Arnstein, *3-D Geometric Origami: Modular Polyhedra*, Dover Publications, 1995.

Rona Gurkewitz and Bennett Arnstein, *Multimodular Origami Polyhedra*, Dover Publications, 2003.

Rona Gurkewitz, Bennett Arnstein, and Lewis Simon, *Modular Origami Polyhedra*, Dover Publications, 1999.

Thomas Hull, *Origami³: Third International Meeting of Origami Science, Mathematics, and Education*, A K Peters, Ltd., 2002.

Thomas Hull, *Project Origami: Activities for Exploring Mathematics*, A K Peters, Ltd., 2006.

Paul Jackson, *Encyclopedia of Origami/Papercraft Techniques*, Headline, 1987.

Kunihiko Kasahara, *Origami for the Connoisseur*, Japan Publications, 1998.

Kunihiko Kasahara, *Origami Omnibus*, Japan Publications, 1998.

Miyuki Kawamura, *Polyhedron Origami for Beginners*, Japan Publications, 2002.

Toshikazu Kawasaki, *Origami Dream World*, Asahipress, 2001 (Japanese).

Toshikazu Kawasaki, *Roses, Origami & Math*, Kodansha America, 2005.

Robert Lang, *Origami Design Secrets: Mathematical Methods for an Ancient Art*, A K Peters, Ltd., 2003.

David Mitchell, *Mathematical Origami: Geometrical Shapes by Paper Folding*, Tarquin, 1997.

David Mitchell, *Paper Crystals: How to Make Enchanting Ornaments*, Water Trade, 2000.

Francis Ow, *Origami Hearts*, Japan Publications, 1996.

David Petty, *Origami 1-2-3*, Sterling, 2002.

David Petty, *Origami A-B-C*, Sterling, 2006.

David Petty, *Origami Wreaths and Rings*, Aitoh, 1998.

Tandalam Sundara Rao, *Geometric Exercises in Paper Folding*, Reprinted by Dover Publications, 1966.

Nick Robinson, *The Encyclopedia Of Origami*, Running Press, 2004.

Florence Temko, *Origami Boxes and More*, Tuttle Publishing, 2004.

Arnold Tubis and Crystal Mills, *Unfolding Mathematics with Origami Boxes*, Key Curriculum Press, 2006.

Makoto Yamaguchi, *Kusudama Ball Origami*, Japan Publications, 1990.

Suggested Websites

- Eric Andersen, *paperfolding.com,* http://www.paperfolding.com/

- Krystyna Burczyk, *Krystyna Burczyk's Origami Page,* http://www1.zetosa.com.pl/~burczyk/origami/index-en.html

- George Hart, *The Pavilion of Polyhedreality,* http://www.georgehart.com/pavilion.html

- Geoline Havener, *Geoline's Origami Gallery,* http://www.geocities.com/jaspacecorp/origami.html

- Thomas Hull, *Tom Hull's Home Page,* http://www.merrimack.edu/~thull/

- Rachel Katz, *Origami with Rachel Katz,* http://www.geocities.com/rachel_katz/

- Hatori Koshiro, *K's Origami,* http://origami.ousaan.com/

- Robert Lang, *Robert J. Lang Origami,* http://www.langorigami.com/

- David Mitchell, *Origami Heaven,* http://www.mizushobai.freeserve.co.uk/

- Meenakshi Mukerji, *Meenakshi's Modular Mania,* http://www.origamee.net/

- Francis Ow, *Francis Ow's Origami Page,* http://web.singnet.com.sg/~owrigami/

- Mette Pederson, *Mette Units,* http://mette.pederson.com/

- David Petty, *Dave's Origami Emporium,* http://www.davidpetty.me.uk/

- Jim Plank, *Jim Plank's Origami Page (Modular),* http://www.cs.utk.edu/~plank/plank/origami/

- Zimuin Puupuu, *I Love Kusudama* (in Japanese), http://puupuu.gozaru.jp/

- Jorge Rezende home page, http://gfm.cii.fc.ul.pt/Members/JR.en.html

- Halina Rosciszewska-Narloch, *Haligami World,* http://www.origami.friko.pl/

- Rosana Shapiro, *Modular Origami,* http://www.ulitka.net/origami/

- Yuri and Katrin Shumakov, *Oriland,* http://www.oriland.com/

- Florence Temko, *Origami,* http://www.bloominxpressions.com/origami.htm

- Helena Verrill, *Origami,* http://www.math.lsu.edu/~verrill/origami/

- Paula Versnick, *Orihouse,* http://www.orihouse.com/

- Wolfram Research, "Origami," *MathWorld,* http://mathworld.wolfram.com/origami.html

- Joseph Wu, *Joseph Wu Origami,* http://www.origami.vancouver.bc.ca/

About the Author

Meenakshi Mukerji (Adhikari) was introduced to origami in her early childhood by her uncle Bireshwar Mukhopadhyay. She rediscovered origami in its modular form as an adult, quite by chance in 1995, when she was living in Pittsburgh, PA. A friend took her to a class taught by Doug Philips, and ever since she has been folding modular origami and displaying it on her very popular website www.origamee.net. She has many designs to her own credit. In 2005, Origami USA presented her with the Florence Temko award for her exceptional contribution to origami.

Meenakshi was born and raised in Kolkata, India. She obtained her BS in electrical engineering at the prestigious Indian Institute of Technology, Kharagpur, and then came to the United States to pursue a master's in computer science at Portland State University in Oregon. She worked in the software industry for more than a decade but is now at home in California with her husband and two sons to enrich their lives and to create her own origami designs. Some people who have provided her with much origami inspiration and encouragement are Rosalinda Sanchez, David Petty, Francis Ow, Rona Gurkewitz, Ravi Apte, Rachel Katz, and many more.

Other Sightings of the Author's Works

◈ *Meenakshi's Modular Mania* (http://www .origamee.net/): A website maintained by the author for the past ten years featuring photo galleries and diagrams of her own works and others' works.

◈ *Model Collection, Bristol Convention 2006,* British Origami Society, diagrams of author's Star Windows was published on CD.

◈ *The Encyclopedia of Origami* by Nick Robinson, Running Press, 2004: A full-page photograph of the author's QRSTUVWXYZ Stars model appeared on page 131.

◈ *Reader's Digest*, June 2004, Australia Edition: A photograph of the author's QRSTUVWXYZ Stars model appeared on page 17.

◈ *Reader's Digest*, June 2004, New Zealand Edition: A photograph of the author's QRSTU-VWXYZ Stars model appeared on page 15.

◈ *Quadrato Magico,* no. 71, August 2003: A publication of Centro Diffusione Origami (an Italian origami society) that published diagrams of the author's Primrose Floral Ball on page 56.

◈ *Dave's Origami Emporium* (http://members.aol. com/ukpetd/): A website by David Petty that features the author's Planar Series diagrams in the Special Guests section (May–August 2003).

◈ *Scaffold*, vol. 1, no. 3, April 2000: Diagrams of the author's Thatch Cube model appeared on page 4.

T - #1033 - 101024 - C92 - 279/216/5 - PB - 9781568813165 - Gloss Lamination